Progress in IS

Progress in IS encompasses the various areas of Information Systems in theory and practice, presenting cutting-edge advances in the field. It is aimed especially at researchers, doctoral students, and advanced practitioners. The series features both research monographs, edited volumes, and conference proceedings that make substantial contributions to our state of knowledge and handbooks and other edited volumes, in which a team of experts is organized by one or more leading authorities to write individual chapters on various aspects of the topic. Individual volumes in this series are supported by a minimum of two external reviews.

Jorge Marx Gómez • Anael Elikana Sam •
Devotha Godfrey Nyambo
Editors

Artificial Intelligence Tools and Applications in Embedded and Mobile Systems

Selected Papers from the First International Conference on Embedded and Mobile Systems (ICTA-EMOS), 24-25 November 2022, Arusha, Tanzania

Springer

Editors
Jorge Marx Gómez
University of Oldenburg
Oldenburg, Germany

Anael Elikana Sam
The Nelson Mandela African Institution of Science and Technology
Arusha, Tanzania

Devotha Godfrey Nyambo
The Nelson Mandela African Institution of Science and Technology
Arusha, Tanzania

ISSN 2196-8705 ISSN 2196-8713 (electronic)
Progress in IS
ISBN 978-3-031-56578-6 ISBN 978-3-031-56576-2 (eBook)
https://doi.org/10.1007/978-3-031-56576-2

This Springer imprint is published by the registered company Springer Nature Switzerland AG
The registered company address is: Gewerbestrasse 11, 6330 Cham, Switzerland

Acknowledgement

The Organizing Committee of the 1st International Conference on Technological Advancement in Embedded and Mobile Systems (1st ICTA-EMoS) 2022 thanks all the implementing partners of the dSkills@EA Project for support and collaboration to host the conference.

EAC
JUMUIYA YA AFRIKA MASHARIKI

INTER-UNIVERSITY COUNCIL FOR EAST AFRICA
IUCEA
QUALITY EDUCATION FOR DEVELOPMENT

german
cooperation
DEUTSCHE ZUSAMMENARBEIT

DAAD
Deutscher Akademischer Austauschdienst
German Academic Exchange Service

giz
Deutsche Gesellschaft
für Internationale
Zusammenarbeit (GIZ) GmbH

Contents

Improving Robustness of Optimized Parameters Gradient Tree Boosting for Crime Forecast Model

Godfrey N. Justo, Leonard P. Binamungu, and Festo S. Clemence

1 Introduction

1.1 Background

Crime is among the common worldwide social problems. It affects lives in different aspects such as interfering with quality of life, economic growth, and national reputation [1]. The global statistics show that there has been an enormous increase in crime rate attributable to internal (self-motivation) and other external influencing factors such as rising unemployment [2] and social inequality. The security agencies especially the police and other departments responsible for public security spend significant time and resources to investigate the trends of crime rate and manner of adapting policing techniques to curb growing trends [2, 3]. In most countries, the police departments still rely on conventional methods such as instinct, experience, evidence, and information collected from the witnesses. Such measures are taken once the crime has been committed [4].

In an attempt to curb the growing crime rate, different countries have been employing several methods such as patrols, public awareness, training, community policing, and crime forecasting. Crime forecasting is advantageous as it can prevent reoccurring of crimes in specific areas or regions by analyzing the patterns of the occurrence of past crimes, which helps in allocating appropriate resources for better police coverage [4]. This relatively new technology enables police departments to access growing volumes of crime-related data that can be analyzed to understand patterns and trends [5]. Several types of crime forecasting models have been pro-

G. N. Justo (✉) · L. P. Binamungu · F. S. Clemence
University of Dar es Salaam, College of Information and Communication Technologies, Dar es Salaam, Tanzania

J. Marx Gómez et al. (eds.), *Artificial Intelligence Tools and Applications in Embedded and Mobile Systems*, Progress in IS,
https://doi.org/10.1007/978-3-031-56576-2_1

posed including statistical models, artificial intelligence models, or hybrid models that are a mixture of both. Commonly used methods for statistical modeling are the Epidemic-Type-Aftershock-Sequence (ETAS) model, Auto-Regressive Integrated Moving Average (ARIMA), linear regression, exponential smoothing, K-means algorithm, Ensemble Poisson-Gamma filter (EnPGF), and Bayesian methods [6]. For AI, Artificial Neural Networks (ANN), support vector regression (SVR), random forest (RF), and Gradient Tree Boosting (GTB) are commonly used [7]. The recent research trends have shifted toward AI models due to their flexibility in handling variations in crime data structure and capability to capture both linear and nonlinear patterns of crime data in which statistical models fail to achieve [7]. Gradient Tree Boosting (GTB) is the most recently applied forecasting AI technique, particularly in crime events analysis. GTB possesses a unique feature among other AI techniques, which is its robustness toward any data representation and distribution. However, the GTB's overall performance is heavily influenced by its input parameter configuration and loss function [7]. The loss function plays a critical role in influencing predictive capabilities and performance by calculating the pseudo-responses value [8].

Primarily, the GTB addresses the parameter configurations challenge by optimizing the main three of its several parameters: namely the number of trees, size of individual trees, and learning rate by using the Dragonfly algorithm (DA), hence the DA-GTB technique, but the loss function in DA-GTB still affects its performance [7]. Indeed, the Dragonfly algorithm with Gradient tree boosting (DA-GTB) has emerged as a popular AI technique for crime forecasting attracting different researchers due to its predictive capabilities compared with other AI and statistical methods [9]. This paper adapts the optimized parameters DA-GTB technique for crime forecast, by determining an optimum loss function. The overall process of the DA-GTB technique applies loss function which plays a critical role in determining the pseudo-response value that in turn affects the predictive capabilities. While a similar study in [8] applied the LAD which is the default loss function for DA-GTB, this study postulates the need to determine a pseudo-response value optimum (minimizing) loss function to replace the DA-GTB default (standard) loss function. The research postulate builds on the role and limitations of loss functions in DA-GTB and the existence of different potential loss functions that can be adopted in the quest for predictive capability improvement. The loss function is the method of evaluating the manner in which an algorithm models the dataset. If the predictions are totally off, the loss function output will be big in number and, if it is pretty good, the output will be small in number. Therefore, the loss function tells us if it is getting somewhere [10].

1.2 Materials and Methods

The DA-GTB technique was used as the main learning method for model development which constitutes the dragonfly algorithm (DA) for optimization of three main parameters of the Gradient tree boosting (GTB) algorithm. In [11], they provide a comprehensive survey of loss functions in machine learning, including a partition criterion of loss functions used in traditional machine learning and deep learning and systematically analyze them. A total of six loss function formulas and corresponding algorithms or applications of common loss functions were designated as suitable for regression. An analysis under the aspects of derivative and robustness, reveals that the Absolute loss, quantile loss, and ε-insensitivity loss are not smooth, but Huber loss is the first derivative and square loss and log-cosh loss are second derivative while the Huber loss, log-cosh loss and absolute loss are robust to outliers. The GTB algorithm is applicable for both regression and classification problems, while regression helps predict a continuous quantity, classification predicts discrete class labels. The crime forecast problem is modeled as a regression problem since the main objective is to forecast the crime rate values as continuous quantities for the different types of crime datasets. Therefore, the analysis of loss functions given in [11] is adapted to identify candidate loss functions that are most promising for optimum results for the forecast problem. In particular, of the six loss functions suitable for regression in [11], the log-cosh loss is dropped due to the problem of gradient and Hessian in addition to the ε-insensitive loss function which does not punish the samples whose error does not exceed ε, making the model focus on the samples with large prediction error.

The forecast models were based on historical crime events data obtained from the Tanzania Police Force (TPF), covering a period of 10 consecutive years from 2010 to 2019. The considered offense/crime data is described in Table 1. Since time is considered the main factor in crime rate forecasting [12], the obtained daily crime event data were used to forecast the crime without considering other features such as population, immigrants, economics, sex, and race [13].

Based on the collected dataset experimental models were developed for each of the four identified loss functions to determine the optimum. The MATLAB software was used for the development of dragonfly algorithm whereas the Python-Scikit-learn tool was used for the GTB and model evaluation. The model evaluation used two metrics, the Root Mean Square Error (RMSE) and Mean Absolute Error (MAE)

Table 1 Criminal offences categories

Crime category	Crime type	Number of events
Theft of arms, theft of motorcycles, theft of motor vehicles, stock theft, theft in a bank, theft in parastatal organizations, theft in cooperative unions, theft in local government, theft in central government, theft in political parties	Theft	519,203
Robbery in highway, armed robbery, robbery with violence	Robbery	88,924
Property breaking	Burglary	300,933

for performance measurement of the output obtained from each experimented loss function on the DA-GTB model as expressed by Eqs. 1 and 2:

$$\mathrm{RMSE} = \sqrt{\frac{1}{n}\sum\nolimits_{t=1}^{n}\left(y_t - \hat{y}_t\right)^2} \tag{1}$$

$$\mathrm{MAE} = \frac{1}{n}\sum_{t=1}^{n}\left|y_t - \hat{y}_t\right| \tag{2}$$

where, $\boldsymbol{y_t}$ is the actual value, $\hat{\boldsymbol{y}}_t$ is the forecast value, and $\boldsymbol{n}$ denotes the number of observations.

The crime forecast model construction process involved four main stages. The dataset was pre-processed to obtain relevant input predictor features using wrapper features elimination method in MATLAB for DA-GTB modelling. The GTB algorithm was configured for its key parameters to guide a weak learner to improve its predictive accuracy through adaptive model decision and implemented based on ensemble machine learning technique which develops a prediction model with boosting and decision tree learning techniques that use multiple base learners to obtain better predictive accuracy. The model applied the different candidate loss functions to determine respective minimization of the pseudo-response value (error fitting) over the response value (crime dataset).

2 Results

2.1 Data Preparation

The data were pre-processed using the wrapper predictive features elimination (PFE) method [14] in order to choose input predictor for easy model interpretation, faster learning time, reducing dimensionality, and reducing over fitting. Based on previous days' corresponding feature analysis it is observed that each sample of nine previous days, the first, second, and third day responded positively for all crime types except for theft crime, which responded positively on the seventh day. Table 2 shows respective selected input predictor features (true) by crime type.

Table 2 Corresponding input predictor features

	Previous days corresponding features								
Crime type	Day 1	Day 2	Day 3	Day 4	Day 5	Day 6	Day 7	Day 8	Day 9
Robbery	True	True	True	False	False	False	False	False	False
Burglary	True	True	True	False	False	False	False	False	False
Theft	True	True	False	False	False	False	True	False	False

Table 3 Optimized parameter values of GTB for each crime type

	Optimized values		
Crime type	Number of trees	Size of individual tree	Learning rate
Robbery	59	8	0.0612
Burglary	68	6	0.0614
Theft	52	10	0.1194

Table 4 DA-GTB with candidate loss functions quantitative measurement error

Crime type	Evaluation metric	LS	LAD	Huber	Quantile
Robbery	MAE	100	36.8	0	48.5
[Huber($\alpha = 0.15$)]	RMSE	100	72.1	0	49.4
Burglary	MAE	100	21	0	38.9
[Huber($\alpha = 0.099$)]	RMSE	100	17.7	0	6.6
Theft	MAE	100	5.5	0	40.1
[Huber($\alpha = 0.11$)]	RMSE	100	5.5	0	29.2

2.2 *GTB Parameter Optimization*

The DA optimal parameter results were obtained through MATLAB by optimizing three main GTB parameters as presented in Table 3. The results show that the minimum optimal number of trees is 52 for theft and the maximum number is 68 for burglary. The findings are attributed to the input predictor features as explained in the data preparation section. With regard to the size of individual trees, the minimum number is 6 for burglary while the maximum number is 10 for theft. The constant number of leaning rates is within the range of 0 and 0.5 while the optimal value is obtained in the range of 0.0612 and 0.1194.

2.3 *Analysis of Candidate Loss Functions*

For analysis of candidate loss functions, DA-GTB experimental models were constructed using crime dataset that comprised three types of crimes: robbery, burglary, and theft; covering a period of 6 years from 2010 to 2015, where 909,060 events were used to train the DA-GTB models as per Table 1. The dataset was fed into the network in order to generate the predictions and compare them with the actual values (the targets) in an attempt to compute the loss (error). The forecast results for each developed model with respective loss function were used to evaluate the performance of the loss function. The quantitative measurement errors of respective developed models are presented in Table 4, where data point x_i is normalized by the range of 0–100 into z_i, such that: [$z_i = x_i - \min(x) / (\max(x) - \min(x)) * 100$].

Table 5 DA-GTB with standard loss function (LAD) and Huber loss function (Huber) quantitative measurement error

Crime type	Evaluation Metric	LAD	Huber	%-Improvement
Robbery	MAE	0.232062	0.222112	4.288%
[Huber($\alpha = 0.15$)]	RMSE	0.305817	0.296627	3.005%
Burglary	MAE	0.163283	0.161753	0.93702%
[Huber($\alpha = 0.09$)]	RMSE	0.21059	0.209446	0.5432%
Theft	MAE	0.191313	0.189723	0.8310%
[Huber($\alpha = 0.15$)]	RMSE	0.261528	0.256001	2.1133%

As observed from the results in Table 4, the choice of the Huber loss function as a pseudo-response value minimizing loss function was attributed to its lowest produced errors as opposed to other candidate loss functions. Thus, the Huber loss function was designated for the implementation of the final improved DA-GTB model, thusly compared to the standard DA-GTB model, which uses standard loss function.

2.4 *Model Comparative Analysis*

Based on the determined optimum loss function, the final improved model employed the Huber loss function. Thus, for comparative purposes the improved model (using Huber loss function) and the standard model (using standard loss function) were developed using three types of crimes dataset; the robbery, burglary, and theft; covering a period of 10 years. The models were evaluated as shown by model's performance on MAE and RMSE metrics for each respective crime type in Table 5. This entails an average percentage improvement of 2.02% and 1.89% for the MAE and RMSE, respectively.

The results show that the Huber loss function with optimal alpha value outperforms other candidate loss functions including the standard DA-GTB least absolute deviation (LAD) loss function. Overall, the crime forecast model with Huber loss function improved the prediction capability compared to other loss functions, in particular, the standard LAD loss function.

3 Discussion

The results from MAE and RMSE for the developed models show that the model that applies the Huber loss function with optimal alpha parameters outperformed the model that applies the standard LAD loss function with optimal alpha parameters in all considered crime types. This is attributed to its lowest MAE and RMSE calculated results. This finding validates the postulate that the DA-GTB crime forecasting model achieves improved predictive capability when implemented by Huber loss function with optimal alpha parameter resulting in minimized pseudo-response value of the model. The process of finding and setting the alpha parameter in the algorithm was done manually by considering its range limit, which is between 0 and 1. The setting starts with the default parameter and then spans through increasing and decreasing the parameter by 0.1 to find the direction that affects the model. After the obtained direction, the value was iteratively used to decrease or increase the alpha parameter until the optimal value was reached. The optimal value for improved models obtained was 0.15, 0.099, and 0.11 for robbery, burglary, and theft, respectively.

Other related studies such as in [8] used the GTB with different loss functions whereby LAD outperformed other loss functions including the Huber loss function with a very small margin. According to [8], the Huber loss function was implemented with its default alpha parameter ($\alpha = 0.5$) for all datasets regardless of their outlier sensitivity. The study on parameter optimization of gradient tree boosting using dragonfly algorithm in crime forecasting analysis in [7] also provided a wide-ranging survey of loss functions used in machine learning by [11]. Both studies proposed need for selection of optimum loss function specific to the problem in order to achieve robustness; but gave no attention to finding both the optimum loss function and associated optimal alpha parameters that attempt to control the outlier in the datasets which improves performance robustness of DA-GTB model. Unlike the studies in [7, 8], this study leveraged the standard advantages of Huber loss function, and in addition, sought to find the optimal alpha parameters for specific datasets in an attempt to control the outlier.

4 Conclusion

Understanding the behaviors of the crime trends is vital for crime prevention through crime forecasting. The application of GTB requires optimization of its parameters, hence the DA-GTB lends high performance compared to GTB alone. The GTB also applies a stage-wise additive framework which improves the learner from the previous weak learner. The overall process of the DA-GTB technique applies loss function which plays a critical role in determining the pseudo-response value that in turn affects the predictive capabilities.

This paper examined different loss functions to enhance crime forecasting performance of DA-GTB-based model. The key research finding advocates the use of Huber loss function in addition to the selection of appropriate optimal alpha parameter. This was achieved through iterative training and testing of the model based on crime dataset types. An important limitation of our results is that the optimum is dependent on specific training data pattern, moreover, other important factors/features such as geo-location, economics, etc., were not considered which are essential in practice as can aid crime planning and allocation of prevention measures. Further research can be directed toward finding efficient methods for determining the optimal alpha value of optimum loss function for model construction efficiency and also consider other prime crime factors.

Acknowledgments The authors are grateful to UDSM for providing supportive research environment and to the TPF for granting permission for use of its data for this research.

References

1. Kaur, N. (2016). Data mining techniques used in crime analysis. *A Review - International Research Journal of Engineering and Technology, 3*(8), 1981–1984.
2. Mittal, M., Goyal, M., Sethi, K., & Hemanth, J. (2019). Monitoring the impact of economic crisis on crime in India using machine learning. *Journal of Computational Economics, 53*(4), 1467–1485.
3. Stec, A., & Klabjan, D. (2018). *Forecasting crime with deep learning* (pp. 1–20). Cornell University.
4. Khairuddin, R., Alwee, R., & Haron, H. (2019). A review on applied statistical and artificial intelligence techniques in crime forecasting. *IOP Conference Series: Materials Science and Engineering, 551*, 012030.
5. Catlett, C., Cesario, E., Talia, D., & Vinci, A. (2018). A data-driven approach for spatio-Temporal crime predictions in smart cities. In *Proceedings-IEEE International Conference on Smart Computing, SMARTCOMP* (pp. 17–24). IEEE. https://doi.org/10.1109/SMARTCOMP.2018.00069
6. Kumar, M., Athulya, S., Mary, M., Vidya, V., Aiswaria, L., Anjana, S., & Manojkumar, K. (2018). Forecasting of annual crime rate in India: A case study. In *International Conference on Advances in Computing, Communications and Informatics, ICACCI* (pp. 2087–2092). IEEE. https://doi.org/10.1109/ICACCI.2018.8554422
7. Khairuddin, R., Ali, A., Alwee, R., Haron, H., & Zain, M. (2019). Parameter optimization of gradient tree boosting using dragonfly algorithm in crime forecasting and analysis. *Journal of Computer Science, 15*(8), 1085–1096.
8. Khairuddin, R., Alwee, R., & Haron, H. (2020). A proposed gradient tree boosting with different loss function in crime forecasting and analysis. *Advances in Intelligent Systems and Computing, 1073*, 189–198.
9. Nair, S., Soniminde, S., Sureshbabu, S., Tamhankar, A., & Kulkarni, S. (2019). Assist crime prevention using machine learning. *SSRN Electronic Journal*, 1–5.
10. Cheng, D., Gong, Y., Zhou, S., Wang, J., & Zheng, N. (2016). Person re-identification by multi-channel parts-based CNN with improved triplet loss function. In *Proceedings of the IEEE Computer Society Conference on Computer Vision and Pattern Recognition* (pp. 1335–1344). https://doi.org/10.1109/CVPR
11. Wang, Q., Ma, Y., Zhao, K., & Tian, Y. (2020). A comprehensive survey of loss functions in machine learning. *Annals of Data Science, 9*, 187–212.
12. Yuki, J., Mahfil Quader Sakib, M., Zamal, Z., Habibullah, K., & Das, A. (2019). Predicting crime using time and location data. In *ACM International Conference Proceeding Series* (pp. 124–128). https://doi.org/10.1145/3348445.3348483
13. Yerpude, P., & Gudur, V. (2017). Predictive modelling of crime dataset using data mining. *International Journal of Data Mining & Knowledge Management Process (IJDKP), 7*(4), 43–58. https://doi.org/10.5121/ijdkp.2017.7404
14. Waad, B. (2015). *On feature selection methods for credit scoring*. PhD Thesis. https://doi.org/10.13140/2.1.3354.1443

Evidence-Based Practices on Co-operative Societies Information Record Management

George Sizya Germinous and Mussa Ally Dida

1 Introduction

Co-operative society is a group of people residing in the same village or group of villages, who pool funds through regular contributions, with the objective of then providing loans to members [1]. It has proved to be one of the driving forces in the socio-economic empowerment of its members [2]. Co-operative societies are established to deal with production, purchasing, processing, marketing, distribution, and any such activity provided for in the rules and by-laws; Nevertheless, co-operative societies involve; savings and credit co-operative societies, financial co-operatives, Consumer co-operatives, housing co-operatives, Industrial co-operatives, Mining co-operatives, microfinance institutions, and co-operative banks which focus on mobilizing savings and providing credit services to households, smallholder producers, and micro-enterprises in both rural and urban areas.

Globally, co-operatives can be sorted out in hierarchical structures comprising primary societies, secondary societies, apexes, and federations. Co-operative structure may differ depending on country's rules and by-laws and membership nature. A case in Kenya, co-operative structure may include an apex society to oversee activities performed by unions of different structures which depends on their membership nature. In Tanzania, co-operatives have been an important part of the economic development. The Government of Tanzania has been implementing the

G. S. Germinous · M. A. Dida
School of Computational & Communication Science and Engineering (CoCSE), Nelson Mandela African Institution of Science and Technology (NM-AIST), Arusha, Tanzania

G. S. Germinous (✉)
Department of Information and Communication Technology, Moshi Co-operative University (MoCU), Moshi, Tanzania
e-mail: georgeg@nm-aist.ac.tz

J. Marx Gómez et al. (eds.), *Artificial Intelligence Tools and Applications in Embedded and Mobile Systems*, Progress in IS,
https://doi.org/10.1007/978-3-031-56576-2_2

Poverty Reduction Strategy by encouraging people to form co-operatives in order to improve their economic prospects [3].

The establishment of the primary co-operative society involves a process that passes through district co-operative office, regional co-operative office, and the registrar of co-operatives' office at the national level. These processes bring about the issue of documentation and records keeping. The information and reports regarding co-operative societies are solely availed at the district co-operative office and not openly available to the public [4]. Bulkiness of the records and procedural requirements to transfer data from one level to another brings difficulties in incorporating all the recorded data regarding a specific primary co-operative society at the district level when moving to either regional or national level during the registration process [5]. Nevertheless, co-operative records are being kept in hard file cabinets, excel sheet forms, and other auxiliary storage devices like memory sticks and external hard disks but the questions remain; are they enough for record management, especially when it comes to bulkiness of records, data redundancy, time consumption in accessing records, and durability? are the records well updated? what about reliability issue during record accessing? On the other hand, there is a limited concentration on keeping records and making them available to the public.

Therefore, this paper presents the evidence-based practice of co-operative record management which is the result of data collection and analysis conducted in the Kilimanjaro region, Tanzania. In addition, this paper provides a technologically designed solution based on web-mobile application that will; improve dissemination of information regarding primary co-operative societies' statistics in the country, thus minimizing the cost needed to acquire such information, reduce or split the power of custodians of the related information, encouraging gender equality and membership openness in primary co-operatives, and lastly, encourage researchers to improve the co-operative sector due to the easy way of accessing related data. Other parts of the document are organized as follows: Sect. 2 introduces a review of associated studies in the literature review. Section 3 focuses on the description of study area, sample size and the technique for sampling, the tools and techniques that guided this study, and the way data were analyzed. Section 4 displays the results of the study as well as discussion of the reported results.

2 Literature Review

The development of co-operative movement has its roots in the far-reaching economic social and political changes which took place in Europe in the late eighteenth and all through the nineteenth centuries, especially the industrial revolution and the liberation of peasantry from former feudalistic system of life [6]. In Western countries particularly in England, it marked the first consumer co-operative society established by men like Robert Owen (1777–1856) and Dr. William King through their theories and efforts.

The success of co-operative movement in Tanzania in the development process marked the raised interest of the government by initiating and establishing various co-operative supporting institutions. Among the established and existing supporting institutions are Moshi Co-operative University (MoCU) and Tanzania Co-operative Development Commission (TCDC). MoCU, then called Co-operative College Moshi (CCM), provides co-operative education and training, research, and consultancy, especially in the co-operative field. The MoCU has offices and co-dominators in all regions in Tanzania with the aim of providing close services to co-operatives throughout the country.

Recently, Information Communication Technology (ICT) has enhanced communication, reporting, and processing of information or records electronically. It may also, facilitate management of co-operatives by enhancing management of records, reports, financial information, and members' management [7]. Moreover, the management of transactions contributed by members of co-operative societies is well managed by the use of ICTs as it reduces the time used in processing payments and retrieving financial statements [8].

Co-operatives with information systems provide a way for storing data and exchanging data among regions. It has been stated that in order to provide viable co-operatives information system solutions, three facets must be considered; Organizational facet, group collaboration facet, and systems facet. Organizational facet concerns with how organization manages its business process in a formal way regardless of who is performing the work (group collaboration) and what technology is used (system facet), group collaboration involves how people work in a common business process and system facet involves the use of different technologies to meet organization needs [9]. The researcher stated that the use of a combination of three facets for co-operative societies enhances performance. Moreover, the researcher insists on the use of co-operative information system to process the data and store records in order to increase accuracy and efficiency [9].

3 Methodology

3.1 Description of the Study Area

The case for this study was Kilimanjaro Region with three district co-operative officers of Moshi municipal, Moshi rural, and Rombo, and 13 respondents from MoCU's regional offices representing the general public; The Kilimanjaro region was selected among Tanzania regions due to the availability of a large number of co-operative societies compared to other regions and presence of three key stakeholders of the study; District Co-operative Officers (DCOs), Researchers (who represented the public), and Registrar of co-operatives.

3.2 Sample Size and Sampling Technique

The study involved 20 respondents whereas, three respondents were district co-operative officers of Moshi municipal, Moshi rural and Rombo, 13 respondents from MoCU's regional offices who represented researchers as well as the general public. Other stakeholders involved were the officers from the Registrar's Office.

3.3 Data Collection Method

During the study, data were collected in the period of 2 months, January and February. The study used questionnaires, interviews, observations, and document review as tools for data collection. Questionnaires were distributed to 13 respondents who were researchers in the co-operative and ICT arena to gather information on current process for acquiring co-operative records. Semi-structured interviews were used to key informants whereby three District Co-operative Officers (DCOs) as well as four representatives from the office of the Registrar of co-operatives (wherein refers to Registrar) were involved. The objective of the interview was to understand the manual process of recording and preparing co-operative societies reports. The observation was conducted physically at the DCO Office observing how employees used the manual process of keeping the co-operatives' records. The work observed the use of papers and ink to update co-operative societies' information and found errors in recording some of the financial calculations. In addition, the study observed how the employees compile data from different co-operative societies and found that the process of compiling data takes a long time. The document review method in this study was to review the reports prepared by the co-operative societies, DCOs, Deputy Registrar of Co-operatives (DRCOs), and the Registrar. In addition, forms used by the DCOs to register co-operatives societies were reviewed.

3.4 Data Analysis Method

The shared out study questionnaires had both hard copy and soft copy structured questions which were managed by using the Open Data Kit (ODK) software. Aftermath, the gathered statistical data were analyzed using a mixed approach whereby a quantitative approach was used with the purpose of; firstly, knowing existing ways for accessing and keeping co-operative records as well as the use of co-operative records being searched, reliability of co-operative information being provided, understanding the challenges experienced in the existing way of handling co-operative data, identifying respondents' experience in using online resources as well as tools being used, and finally, to capture proposed information that should be in the CRMS and reports to be generated. On the other hand, quantitative approach was used with the aim of; identifying tools that are in use for processing co-operative data, grasping the extent of challenges experienced when acquiring co-operative information, and obtaining suggestions on the ways of improving the existing process of accessing co-operative information.

In all above-mentioned cases, SPSS software as a statistical package was used for analyses. The reason being, SPSS provides the easiest and customary virtualization of tables and charts compared to other statistical packages like Excel, SAS, and R. For example, Excel requires formulas for every calculation to be entered manually thus providing room for errors. In conjunction with that, R and SAS are thorny to learn, conditional to programming locale [10].

4 Results and Discussion

4.1 Tools Being Used for Processing Co-operative Data

The study aimed at knowing tools being used by DCOs and the Registrar's office to process co-operative data which they normally receive from the field. All respondents mentioned that the processing of co-operative societies' records is normally conducted by using manual tools such as "pens and papers," and electronic tools such as Excel sheets and word processors, whereas memory sticks and external hard disks were used as storage devices.

4.2 Existing Ways for Accessing and Keeping Co-operative Records

The study sought to know how the public access co-operative information. Respondents mentioned only two ways; "either by physically paying a visit to respective offices for detailed information or by going through the TCDC website for summarized reports" as it may be observed in Table 1.

Table 1 Existing record acquisition methods

Methods	Frequency	Percentage
Physically visiting respective offices	13	100
Using online resources	9	69.2
Asking others (colleagues)	7	53.2
Reading books and other articles	3	22.1
Other	0	0

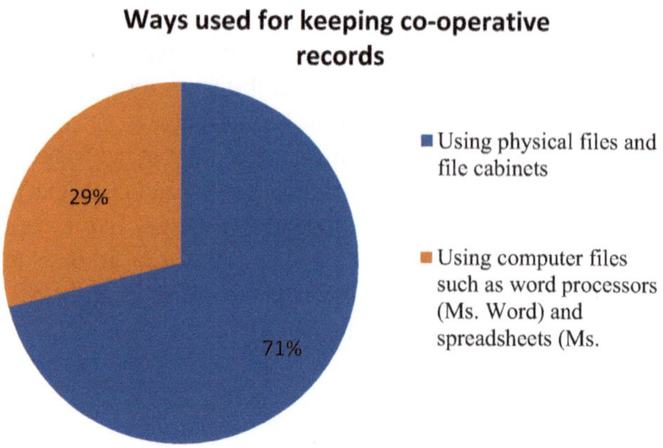

Fig. 1 Presents ways being used in keeping co-operative records

Table 2 Usage of co-operative information being searched

Use	Frequency	Percentage
Research purpose	13	100
For policy issues	7	53.9
Out of curiosity	6	46.2
Just for knowledge	11	84.6
Training and consultancy (other)	6	46.2

On the other hand, the researcher interviewed DCOs as well as Registrar's office on various ways that are used for keeping different records about co-operatives as shown in Fig. 1.

4.3 *Use of Co-operative Information Being Searched*

The study aimed to assess public usage of information that is being processed and kept by DCOs as well as Registrar's office. Table 2 presents results from respondents on the usage of co-operative information.

Statistically, all 13 respondents which are equal to 100% proved to use the co-operative data for research purposes. Apart from research, 11 of them pointed out to use of the information for knowledge enrichment, which is equivalent to 84.6% and 7 of them used for policy-related issues which is equivalent to 53.9%. Lastly, six respondents equal to 46.2% revealed to use the information for training and other consultancy issues, while the same frequency normally searches the information just out of curiosity. These results are summarized in Table 2.

4.4 Reliability of Co-operative Information

To know how to obtain the required co-operative data when needed, the study provided YES or NO questions to the respondents. Therefore, it was revealed that, a small number of only four respondents which is equivalent to 33% showed that they normally get what they wanted whenever in need while nine of them equal to 67% normally do not obtain co-operative data as required. Percentage distribution clearly shows unreliability of co-operative data whenever acquired by those in need of them.

4.5 Challenges Experienced in the Existing Process of Handling Co-operative Data

In this part, the study wanted to acquire all the challenges experienced by respondents as a result of using existing methods for processing, keeping, and accessing co-operative information. The study observed that, out of 13 respondents only eight responded to the question whereby all of them agreed that data inaccuracy and inconsistency as well as lack of single source of obtaining required information are the main challenges being experienced in the whole process of acquiring co-operative data. Also, seven respondents which is equivalent to 87.5% had similar opinions on the challenges experienced during acquisition of co-operative records by pointing out the following; bureaucracy during physical access, untimely provision of the needed co-operative information, and provision of un-updated information, inadequate online resources, and most co-operatives do not keep records. Other challenges pointed out with their percentage were; time consumption due to geographical challenges (75%), increased cost in data acquisition (62.5%), lack of transparence (50%), and improper presentation of co-operative information in request (12.5%). The percentage shows that there are many challenges that are faced in existing manual process of co-operative data handling as it may be grasped in Fig. 2.

4.6 Extent of the Challenges in Acquisition of Co-operative Information

The study aims at finding out the extent of challenges in acquiring co-operative information for users. Therefore, the researcher provided a scale to measure extent of the challenges by using numbers 0 to 5 as shown in Table 3 where; 0 meant the challenge was not to a high extent while 5 meant the challenge was in a very high extent. In the series of provided challenges, data inaccuracy and inconsistency were mostly declared to be a very serious problem. This is because out of 13 respondents,

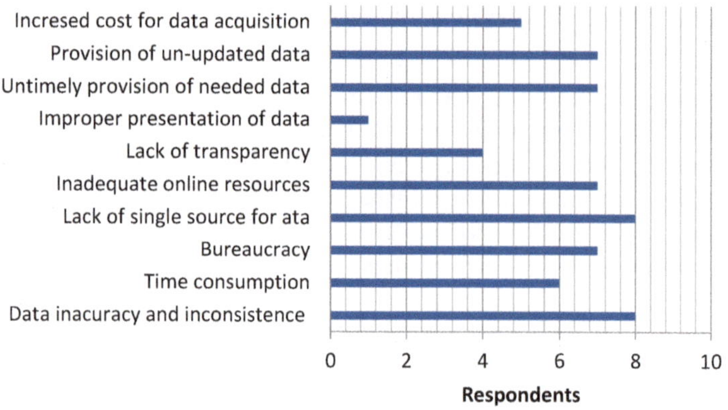

Fig. 2 Presents challenges experienced in the existing process for co-operative data handling

Table 3 Presents the extent of the stated challenges in acquisition of co-operative information

Challenges	0	1	2	3	4	5
Data inaccuracy and inconsistency	0	1	1	1	3	7
Wastage of time during the process	0	1	1	3	3	5
Increased data acquisition cost	1	2	1	8	1	0
Bureaucracy	0	0	1	2	9	1
Lack of transparency	4	2	2	1	2	2
Improper presentation (format) of requested information	1	0	1	1	7	3
Communication barrier	2	4	4	1	1	1

7 of them chose number 5 which meant it was a very serious problem. On the other hand, respondents felt that increased costs in data acquisition were a moderate challenge because 8 out of 13 respondents selected number 3 which meant to be a problem but not that adverse. The study presents a graphical representation of respondent's choices to show the extent of given challenges as shown in Table 3.

4.7 Respondent's Experience on Using Online Resources

The study sought to know respondents' experience on using online resources. It was found that about 46% of respondents were strongly dependent on online resources while 54% of respondents were average consumers of online resources as depicted in Fig. 3. Therefore, all respondents use online resources.

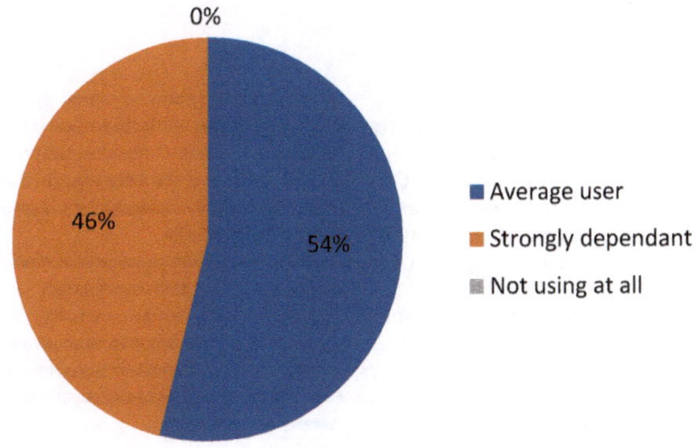

Fig. 3 Presents respondent's experience on using online resources

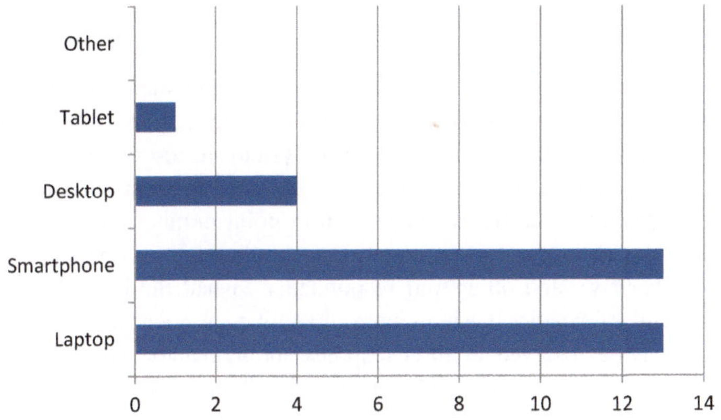

Fig. 4 Presents tools currently being used to access online resources

4.8 Tools Currently Being Used to Access Online Resources

This part explored on tools or devices that are currently being used by respondents in order to access online resources. It was revealed that all 13 respondents which was equivalent to 100% use smart phones and laptops in order to access online resources. Figure 4 shows that out of 13 respondents, 4 of them included desktop computers to access online resources while only 1 respondent used a tablet.

4.9 Suggested Ways of Improving the Existing Process of Accessing Co-operative Data

The study provided an avenue for respondents to suggest better ways to improve the way co-operative records are being accessed or disseminated. From the snipped

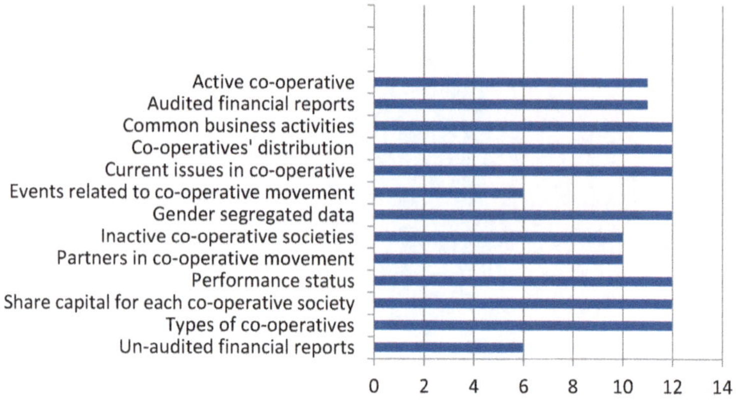

Fig. 5 Represents information proposed to be in CRMS

image of responses below, it was found that some respondents wished for institutions to invest in online resources which actually utilize the use of information technologies through management information systems in co-operatives. A snapshot displayed in Fig. 5 clearly shows that four respondents mentioned the need for more investment in online resources as seen in point number (i), six respondents saw the need for proper use of information technologies in co-operatives as seen in point number (iv), and other four respondents wished the installation and use of management information systems in co-operatives as seen in point number (v) below. As it may be observed, all three responses pointed above had a common view in point, i.e., more investments in ICT. On the other hand, more suggestions insisted on proper handling of co-operative data being kept whereby two respondents saw the need for safe keeping data and information as seen in point number (ii), and the other three suggested frequent updating of co-operative data in point number (iii). Then again, there was a respondent who claimed just to be comfortable with the prevailing process.

4.10 Information Proposed to Be in CRMS

Figure 5 displays the proposed information supposed to be in the management system (CRMS). The study provided an avenue in the questionnaire for respondents to select the type of information they would like to access in the CRMS from the given list.

Fig. 6 Represents reports to be generated by CRMS

4.11 *Reports to Be Generated by CRMS*

The study required respondents to indicate the kind of reports they would like to access from CRMS after processing. Respondents were provided with an anticipated list of reports respondents to make a choice out of them as seen in Fig. 6. The list allowed respondents to make more than one choice.

4.12 *Conclusion and Future Work*

Based on the study findings about the challenges faced by users on the existing manual process of co-operatives registration, reporting, and recording management; this study proposed a solution which is Co-operative Record Management System (CRMS) to solve challenges. The CRMS provides a solution that will enable DCO's Office to record; manipulate and generate reports on co-operative societies' data electronically and reduce challenges concerning with time, error in recording financial details, and loss of information. The flow of the proposed solution starts from co-operative societies, DCOs, DRCOs, and Registrar. Co-operative societies submit their details to the DCOs Office manually then DCOs Office register and feed the co-operative details to the system then submit to the DRCOs Office. Thereafter, DRCOs Office manipulates data and submits to the registrar. All the offices can search and generate reports according to their requirements. Moreover, all the processes are conducted through the system as seen in Fig. 7.

Finally, regardless of the vivid truth that, the study concentrates on and tries to speak about the challenges in front of current manual ways for handling and managing co-operative records in Tanzania, yet the study provides an opportunity for others to put in their technological incites on internal operations of an indi-

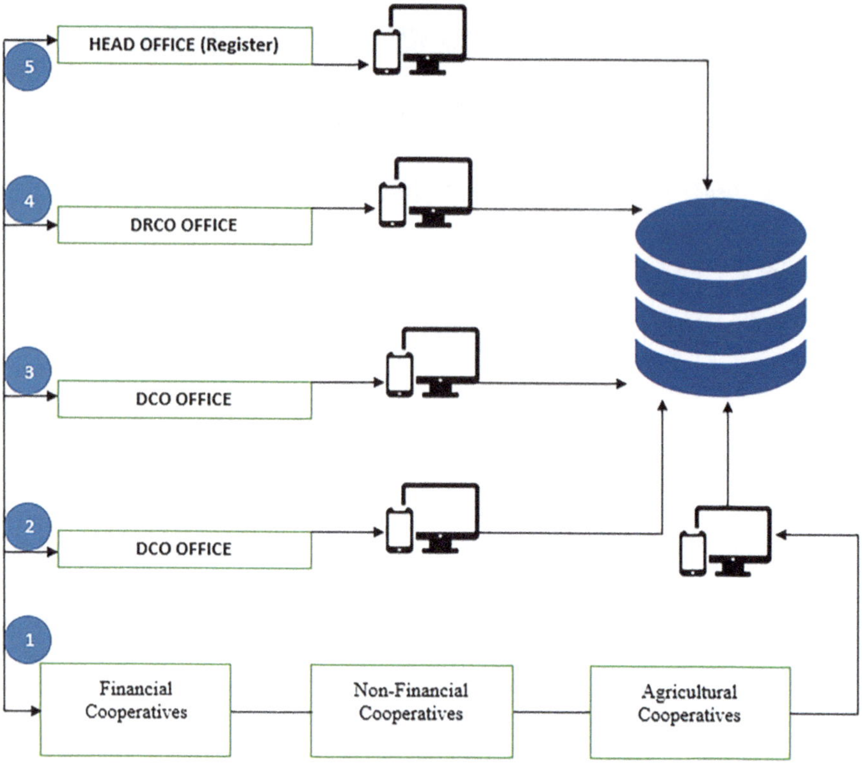

Fig. 7 The general framework of the proposed software solution

vidual co-operative society such as loans and membership management, leadership appointment processes, and preliminary activities conducted before co-operative society's registration process. Also, the study allows future works on the capability for policymakers to forecast the financial abilities of co-operative members in co-operative societies as well as societies to project the membership trend throughout the co-operative industry.

References

1. Maleko, G., & Msuya, R. (2015). Women participation in co-operatives—Challenges and prospects. The case of selected Saccos and Amcos in Kilimanjaro and Arusha regions Tanzania, East Africa. *Journal of Business Administration and Education, 7*(1), 81–111. http://infinitypress.info/index.php/jbae/article/view/950
2. Dogarawa, A. B. (2012). The role of co-operative societies in economic development. *SSRN Electronic Journal*, 1–20. https://doi.org/10.2139/ssrn.1622149
3. United Republic of Tanzania. (2005). United Republic of Tanzania national strategy for growth and reduction of poverty (06), 1–119.

4. Kayunze, K. A., & Kimaryo, L. P. (2020). Members' views on ownership and democratic sustainability in primary agricultural marketing co-operatives in Tanzania; Evidence from 1(January), 71–84.
5. United Republic of Tanzania (URT). (2014). The-Cooperative-Societies-Act-2013.pdf
6. Cronery Rwekaza, G., & Mhihi, B. (2016). Co-operative development in Tanzania: A tool for equality and socio-economic development. *Journal of Economics and Sustainable Development, 7*(6), 29–40. www.iiste.org. Www.Iiste.Org ISSN
7. Maghimbi, S. (2010). *Cooperatives in Tanzania mainland: Revival and growth.* http://www.ilo.org/public/english/employment/ent/coop/africa/info/publ.htm
8. Suyono, E., Rusmana, O., & Riswan, R. (2019). The revitalization model through the implementation of accounting information system for village unit cooperative in Banyumas region, Indonesia. *Media Ekonomi Dan Manajemen, 34*(1), 1–13. https://doi.org/10.24856/mem.v34i1.870
9. Zouari, M. B. C., & Dakhli, S. B. D. (2018). A multi-faceted analysis of knowledge management systems. *Procedia Computer Science, 138*, 646–654. https://doi.org/10.1016/j.procs.2018.10.086
10. Fujo, M. H., & Dida, M. A. (2018). Web-based admission system for advanced level, private schools: Case of Kilimanjaro region, Tanzania. *International Journal of Advanced Technology and Engineering Exploration, 5*(47), 407–418. https://doi.org/10.19101/ijatee.2018.546020

Influence of Microservice Design Patterns for Data Science Workflows

Christoph Schröer, Raphael Holtmann, Jorge Marx Gómez, and Hergen Pargmann

1 Introduction

In general, data science workflows support the automation of data-driven decision processes. Using machine learning approaches based on huge amount of data, data science could improve such processes in several domains. Several architectural styles have evolved like lambda architecture in the field of big data [4] or workflow management systems in the field of business processes. Also microservice architectures could enforce the implementation and deployment of data science workflows [10].

Following the microservice approach, several microservice patterns exist to support the implementation of such architectures in general. For data science workflows, several microservices could be involved. Therefore, challenges are how multiple microservices could be scheduled in the correct order regarding a workflow model and how these communicate with each other in terms of data exchange. If functionalities are divided among several services, they are no longer executed in the same process but distributed among several.

In practice and theory, design patterns for business process execution and communication styles within microservice architectures have evolved. For the implementation of microservice-oriented business processes, orchestrator and choreography patterns exist. For communication, synchronous and asynchronous styles exist [5, 6]. These are extended if a distinction is made between control flow and data exchange and thus further patterns exist for data exchange.

C. Schröer (✉) · R. Holtmann · J. Marx Gómez
University of Oldenburg, Oldenburg, Germany
e-mail: christoph.schroeer@uni-oldenburg.de

H. Pargmann
Jade Hochschule, Oldenburg, Germany

J. Marx Gómez et al. (eds.), *Artificial Intelligence Tools and Applications in Embedded and Mobile Systems*, Progress in IS,
https://doi.org/10.1007/978-3-031-56576-2_3

While these patterns have been well researched for transaction-oriented applications, there is a lack of using these in data science workflows. Thus, this article aims to investigate under which conditions certain design patterns are suitable. We use experimental prototyping of a typical data science workflow including the data preparation, training of a machine learning model and their inference. This prototype is tested under different volume and velocity workload conditions that are typically for data science workflows.

We see both practical and theoretical implications. Practically, software architects could be supported in the implementation of data science workflows with suitable design patterns. Theoretically, we have linked the microservices pattern to data science workflows.

In the next section, we place this paper into the current state of research. In the third chapter, we formulate the research question, present the method and the selected metrics. In chapter 4, the data science workflow and the derived microservice architectures are described. In chapter 5, the results of the experiments are shown. In the last section, the results are discussed and finally the contributions, implications, and outlook are summarized and explained.

2 Related Work

This article addresses the research areas of microservices design patterns and workflow design patterns and the link between them.

For a first overview of microservice patterns in general, there are fundamental books structuring them to messaging, discovery, data consistency or querying pattern [6] or to communication and transaction patterns [5]. They also offer qualitative decision support when choosing the right pattern. Also in recent literature, microservice patterns are structured into patterns for data persistence, communication or entry points [12] or similar in communication, deployment and data storage [11].

Furthermore, various microservice patterns have been studied experimentally and quantitatively. For an e-commerce application, distributed transactions are implemented with orchestration and choreography patterns. The performance of the saga implementation is investigated after rollbacks with the result that orchestration lasts longer than choreography [7]. The saga pattern is a design pattern for designing distributed transactions [6]. Also, the patterns of Application Programming Interface (API) Gateway, chain of responsibility, and asynchronous communication are investigated for transaction-oriented application by the response times under different workloads [1]. During a migration from monolith to microservice architecture, two different microservice architectures are compared. They differ in communication mechanisms and data management. The metrics are the average response times, CPU, and memory requirements by requests per minute. Regarding this metric, synchronous communication without a focus on data consistency performs best [3].

Independent from the architectural choice of microservices there are also workflow patterns proposed in literature. Also, specific design patterns exist for workflows like control-flow patterns, data patterns, resource patterns, and service interaction patterns [9]. However, these have not yet been linked to microservice patterns.

To summarize, the investigation of relevant microservice patterns for data science workflows is lacking here.

3 Research Method

For researching the influence of microservice design patterns on the microservice-oriented implementation of data science workflows, we formulate the following research question:

> **? Research Question**
>
> To what extent can software architects be supported in selecting microservice design patterns for data science workflows?

For addressing the research question, we use the research method of prototyping combined with laboratory experiments [3, 13]. For prototyping, we use a typical workflow and implement it using microservices. The workflow and architecture are presented in the next section. For laboratory experiments, we construct a practical environment using the cloud provider Amazon web services[1] and defining use case related workloads. We use experiments to investigate the impact of architectures and workloads on workflow performance.

For workflow performance, we use the metric of end-to-end delays of a whole workflow execution [2]. The workflow end-to-end-delay WED means the time that a workflow h takes for a whole execution, including all processing functions and infrastructure time. Therefore, Eq. 1 uses all beginning and ending timestamps TSB and TSE of all operations opr within a workflow wf_h that are executed under a specific workflow configuration e, workload λ, and architecture α. Furthermore, $\overline{WED}(\lambda, \alpha)$ is the mean of all end-to-end delays of all workflow configurations and their executions by workload and architecture.

$$WED_h(e, \lambda, \alpha) = \begin{array}{l} max\,(TSE_k\,(e, \lambda, \alpha)\ :\forall opr_k\ in\ wf_h) \\ -\ min\,(TSB_k(e, \lambda, \alpha)\ :\forall opr_k\ in\ wf_h) \end{array} \tag{1}$$

[1] https://aws.amazon.com/ (Last accessed on 11th September 2022).

To compare microservice architectures with each other, the average changes of the end-to-end-delay per workflow execution compared to a baseline workload λ_0 are calculated (see Eq. 2). The mean of the changes is calculated in Eq. 3. The E_λ is defined as the set of all workflow configurations under a specific workload configuration. For example, to test low volume workload, workflows are configured to select only a small amount of data for the workflow to process.

$$\Delta WED(e, \lambda, \alpha) = \frac{WED(e, \lambda, \alpha) - WED(e, \lambda_0, \alpha)}{WED(e, \lambda_0, \alpha)} \tag{2}$$

$$\overline{\Delta WED}(\lambda, \alpha) = \frac{\sum_{\text{all } e \in E_\lambda} \Delta WED(e, \lambda, \alpha)}{|E_\lambda|} \tag{3}$$

4 Data Science Workflows and Microservice Architectures

Data science workflows vary in complexity depending on the use case. For testing suitable design patterns, we focus on typical processing functions within data science, namely training and inference of machine learning models. Therefore, we select three processing functions from a company-relevant use case. The use case is developed in the corporate foresight department of an automotive manufacturer. The use case aims to evaluate startups in an automated way using relevant data selection and application of machine learning algorithms. This assessment is then incorporated into subsequent decision support processes. The operations relevant to the study are the preparation of data, the training of a machine learning model, and the inference of the model. These functions are applied in two workflows: the training workflow and the inference workflow. Figure 1 shows the workflows schematically and the order in which they are executed.

Figure 1 shows one possible microservice architecture with two microservices. One microservice implements the control flow of the workflows. The second microservice implements the processing operations. A data exchange layer exchanges data between the processing operations like the processed data, machine learning model, and inference results. Data is not exchanged directly through the interfaces of the microservices (see also workflow pattern for data interaction between workflow tasks [8]). Figure 1 shows the microservices without dependencies. The communication as well as the implementation of the control flow of the workflows depends on the selected design patterns. In this article, two alternatives are compared:

1. Orchestrator pattern with synchronous communication
2. Choreography pattern with asynchronous communication

In principle, the orchestrator pattern can also be designed with asynchronous communication but it is not pursued further here due to too little distinguishability.

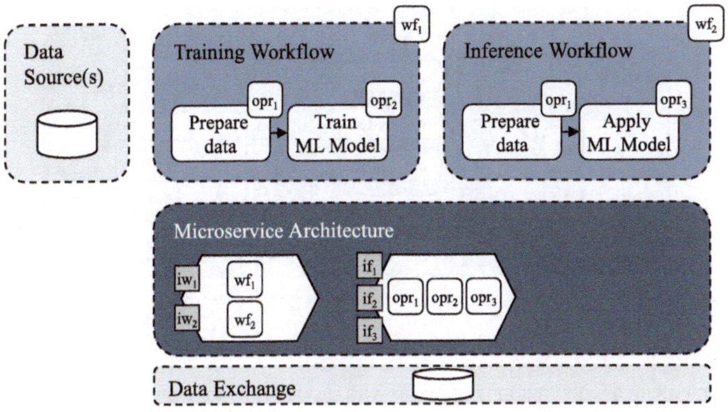

Fig. 1 Investigated data science workflows and corresponding microservices. Gray arrows between data science operations indicate control flow

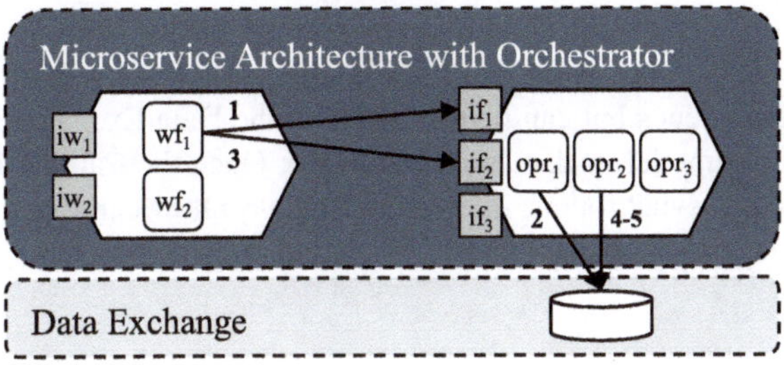

Fig. 2 Implemented orchestrator pattern with synchronous communication. Black arrows between microservices indicate data flow between microservices and data exchange

The choreography pattern usually only takes place with asynchronous communication.

The orchestrator pattern with synchronous communication is shown in Fig. 2 using the training workflow as an example. The workflow is started via the iw_1 interface. The synchronous REST interface if_1 is then called (1). The processed data is not returned as a response but is stored as a file in the data exchange layer (2). The physical location is returned as the response. Then wf_1 calls the interface if_2, which is associated with the operation opr_2 (3). The only parameter is the file name with the data for the training of the machine learning model (4). The machine learning model is also stored in the data exchange layer after training (5). This can then be reused in the inference workflow.

The choreography pattern with asynchronous communication is illustrated in Fig. 3 using the training workflow as an example. The main difference is the use of another infrastructure component of a message broker. The training workflow can be started via the iw_1 interface. The workflow only sends a start message (1). The asynchronous interface af_1 (2) listens for this and starts preparing the data, which is stored in the data exchange layer (3). Then an event is sent that the operation has

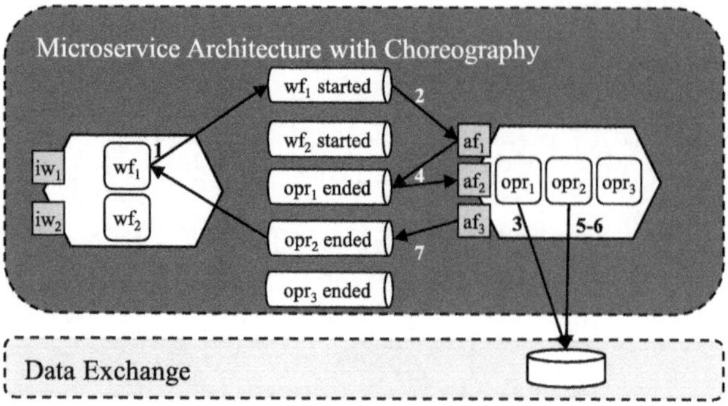

Fig. 3 Implemented choreography pattern with asynchronous communication. Black arrows between microservices indicate data flow between microservices, message queues, and data exchange

finished (4). Since this is the choreography pattern, the asynchronous interface af_2 will listen directly to the event and start training. The data will not be sent along via the message queues but can be retrieved via the Data Exchange Layer (5). The machine learning model is also stored there (6). Once the training is complete, an event is sent. The asynchronous interface of the workflow can listen to this event to know that the workflow execution as a whole has been completed.

5 Experiments and Results

For an experimental investigation of the two design patterns from Figs. 2 and 3, the workflows are subjected to several load tests. There are workflow executions of the training workflow and the inference workflow. In addition, there are multiple workflow configurations, where the workflows are each executed with different independent data selections. A data selection for training means for example that a data filter is set to companies founded between 2014 and 2015 (low volume) or 2000 and 2015 (high volume). In total, this results in 64 data selections for the training workflow and 96 data selections for the inference workflow. These are split into two sets, low data volume (32 training, respectively, 48 inference workflow configurations) and high data volume (32, respectively, 48) workflow configurations. These can be executed sequentially or in parallel. This ultimately leads to the following workloads for the experiments:

- Low Volume with 1, 2, 4, and 32 parallel executions for the training workflow and 1, 3, 8, and 48 parallel executions for the inference workflow (LV-1, LV-2, LV-3, LV-4, respectively). Low volume means an average data volume for each execution of about 14 MB for training and 3 MB for inference.

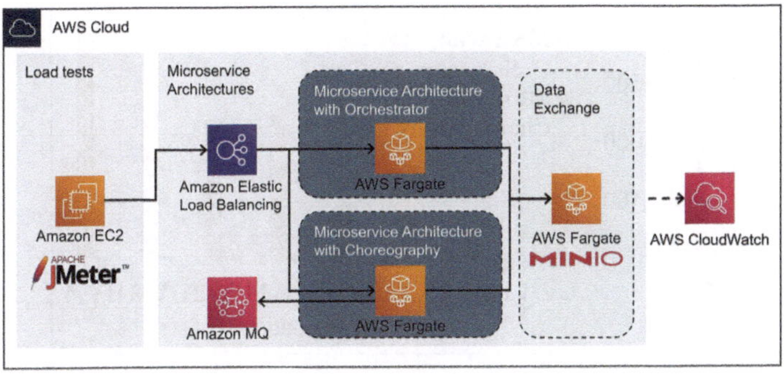

Fig. 4 AWS architecture with microservice architectures and infrastructure components

- High Volume with 1, 2, 4, and 32 parallel executions for the training workflow and 1, 3, 8, and 48 parallel executions for the inference workflow (HV-1, HV-2, HV-3, HV-4, respectively). High volume means an average data volume for each execution of about 46 MB for training and 15 MB for inference.

To compare the differences between the workloads and their effects, we define the baseline workloads LV-1 and HV-1. For example, we compare then the metrics under the workload LV-4 with LV-1 or HV-4 with HV-1. Comparing the metrics between low and high volume configurations does not make sense due to different complexity executions of the underlying algorithms.

The microservices are implemented with Python and deployed on AWS cloud infrastructure as shown in Fig. 4. RabbitMQ[2] provisioned by Amazon MQ[3] serves as a message broker. MinIO[4] serves as a data exchange layer. The workload tests are configured and executed with Jmeter.[5] Jmeter is run on an EC2 instance. The microservices are packaged as docker images and deployed on ECS Fargate instances, each with maximum resources (4 CPU, 30 Gb RAM). Each microservice is deployed as one single instance. Each microservice architecture has its application load balancer.

The results of the experiments are shown in Figs. 5 and 6. Network errors were not observed during the experiments. Figure 5 shows the absolute mean workflow end-to-end delays by workload and by architecture. The microservice architecture with choreography architecture always has higher end-to-end delays. Increasing workload (both volume and velocity) leads to increasing end-to-end delays in all architectures.

Figure 6 plots the average change of end-to-end-delay of workflows to the baseline workloads LV-1 and HV-1, respectively. This makes it clear which architecture

[2] https://www.rabbitmq.com/ (Last accessed on 11th September 2022).

[3] https://aws.amazon.com/en/amazon-mq/ (Last accessed on 23th October 2022).

[4] https://min.io/ (Last accessed on 11th September 2022).

[5] https://jmeter.apache.org/ (Last accessed on 11th September 2022).

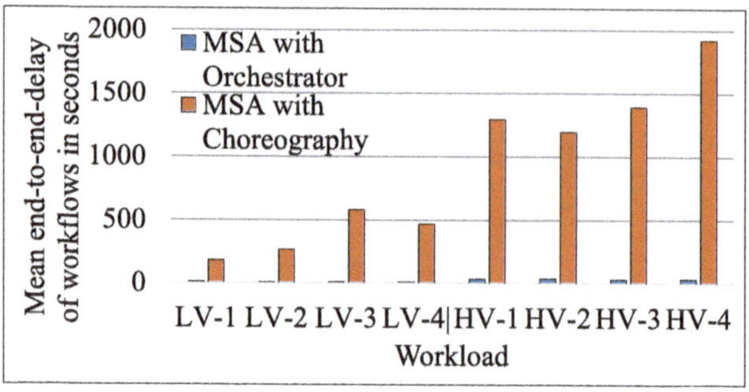

Fig. 5 Mean end-to-end-delay of workflows $\overline{WED}(\lambda, \alpha)$ by workloads and microservice architecture (MSA)

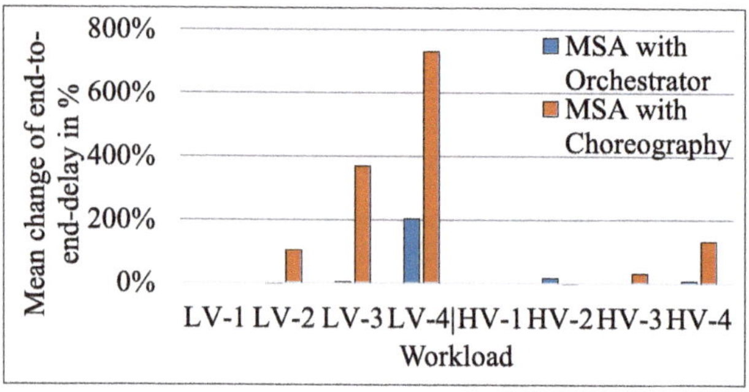

Fig. 6 Mean change of end-to-end-delay of workflows $\overline{\Delta WED}(\lambda, \alpha)$ compared to baseline workloads LV-1 and HV-1 by workloads and microservice architecture (MSA)

can handle increasing workloads (in this case increasing velocity) best. In general, it becomes clear that end-to-end delays also increase as the workload increases. However, this increase differs depending on the architecture. While execution times in the choreography architecture can increase by up to 700% up to workload LV-4, this only increases by up to 200% under the orchestrator architecture.

6 Discussion and Conclusion

The fact that the end-to-end delays of the choreography architecture are higher could be due to the additional infrastructure components of the message broker. These are configured in the asynchronous interface of the microservices for a limited retrieval of messages for a stable connection.[6] This results in longer execution times, especially at higher data velocities.

Since all experiments resulted without any issues, we argue that under the tested conditions, both orchestrator and choreography architecture patterns are suitable. However, the orchestrator pattern has a higher coupling between the microservices by design. If even longer end-to-end delays are expected, the orchestrator pattern will no longer be suitable due to its synchronous communication interfaces. For example, the application load balancer from AWS can be configured with a maximum timeout of 4000 seconds.[7] However, the question remains here to what extent longer execution times can then also be justified for inference. Thus, a combination of both patterns would be possible: the choreography pattern for training and the orchestrator pattern for inference. The end-to-end delays of inference are always lower than those of the training workflow. Thus, one recommendation would be to provide microservices with both a synchronous interface for inference and an asynchronous interface for training. Another option would also be to split the functionality between two microservices.

With these results, we summarize the paper as follows. First, we have linked microservice design patterns to data science workflows. We have shown that both orchestrator and choreography patterns are suitable. In addition, we give an example implementation of how a typical data science workflow can be implemented with microservice design patterns. With the experiments and results, we can also assess of under which workload conditions design patterns perform. Here, we also see that the metric of end-to-end delays is suitable since it takes into account not only the response times of individual interfaces but also those of the entire workflow executions, including the network traffic.

However, the study is faced with some limitations. First, the study is conducted with an concise data science workflow, although data science workflows could be very complex. Second, we have used only one metrics. While this metric is suitable, other metrics could also be evaluated in future. Third, we conducted the experiments in a laboratory where we have examined only a subset of possible factors influencing the architectures. In the future, several factors as well as the influence on the distribution of functionalities across multiple microservices can be investigated.

In summary, we see both practical and theoretical implications. Practically, software architects could be supported in the implementation of data science workflows with suitable design patterns. Theoretically, we have linked the microservices pattern to data science workflows.

---

[6] see also consumer prefetch https://www.rabbitmq.com/consumer-prefetch.html (last accessed on September 11th, 2022).

[7] https://docs.aws.amazon.com/elasticloadbalancing/latest/application/application-load-balancers.html(last accessed on 12th of September 2022).

References

1. Akbulut, A., & Perros, H. G. (2019). Performance analysis of microservice design patterns. *IEEE Internet Computing, 23*(6), 19–27 .
2. Bao, L., Wu, C., Bu, X., Ren, N., & Shen, M. (2019). Performance modeling and workflow scheduling of microservice-based applications in clouds. *IEEE Transactions on Parallel and Distributed Systems, 30*(9), 2114–2129.
3. Costa, L., & Ribeiro, A. N. (2022). Performance evaluation of microservices featuring different implementation patterns. In A. Abraham, N. Gandhi, T. Hanne, T. P. Hong, T. Nogueira Rios & W. Ding (Eds.) *Intelligent systems design and applications* (vol. 418, pp. 165–176). Springer International Publishing.
4. Marz, N., & Warren, J. (2015). *Big data: Principles and best practices of scalable real-time data systems*, 1. aufl. edn. Manning.
5. Newman, S. (2021). *Building microservices: Designing fine-grained systems* (2nd edn.). O'Reilly Media, Inc.
6. Richardson, C. (2019). *Microservice patterns: With examples in Java*. Manning.
7. Rudrabhatla, C. K. (2018). Comparison of event choreography and orchestration techniques in microservice architecture. *International Journal of Advanced Computer Science and Applications, 9*(8), 1–11.
8. Russell, N., ter Hofstede, A., Edmond, D., van der Aalst, W. (2005). Workflow data patterns: Identification, representation and tool support. In L. Delcambre, C. Kop, H. C. Mayr, J. Mylopoulos & O. Pastor (Eds.) *Conceptual modeling - ER 2005, Lecture notes in computer science* (pp. 353–368). Springer.
9. Russell, N., van der Aalst, W. M., & ter Hofstede, A. H. M. (2016). *Workflow patterns*. The MIT Press.
10. Schröer, C. (2021). Towards microservice identification approaches for architecting data science workflows. *Procedia Computer Science, 81*(3), 519–525.
11. Taibi, D., Lenarduzzi, V., & Pahl, C. (2018). Architectural patterns for microservices: a systematic mapping study. *Proceedings of the 8th International Conference on Cloud Computing and Services Science, 1*, 221–232.
12. Valdivia, J. A., Lora-González, A., Limón, X., Cortes-Verdin, K., & Ocharán-Hernández, J. O. (2020). Patterns related to microservice architecture: a multivocal literature review. *Programming and Computer Software, 46*(8), 594–608.
13. Wilde, T., & Hess, T. (2007). Forschungsmethoden der Wirtschaftsinformatik: Eine empirische Untersuchung. *Wirtschaftsinformatik, 49*(4), 280–287.

A Systematic Literature Review Toward Standardization of Business Rules Discovery in the Context of Process Mining

Menna Wael and Gamal Kassem

1 Introduction

Business rules can be seen from a business process perspective as statements or constraints that describe the business process behavior and they reflect what is allowed and what is restricted in the process [1]. Business rules are considered the basic structure of the business process [2] and they are incorporated in different patterns within process data [3]. The discovery of business rules from business processes allows for the analysis, verification, and improvement of the business processes. Discovered business rules help reduce costly mistakes that happen during process execution and help achieve process objectives. The discovery of business rules from business processes can support the detection of process bottlenecks, making process changes, and improving the overall performance of the business process [2]. In addition, discovered business rules are used for compliance checking against intended business process policies or compliance requirements, this helps detect gaps between both intended and executed business rules [4]. Moreover, the discovery of business rules from the process event logs provides a detailed and accurate specification of process behavior [5].

Data mining is used to find hidden patterns [6] and it is a common method to discover business rules from the process event logs, in which different data mining algorithms can be applied to event logs to discover patterns or rules within event log data, that are considered business rules and can be expressed in common business rules formats such as the IF/THEN format or the antecedent/ consequent format [2, 7]. Business rules discovered through the application of data mining algorithms

M. Wael (✉) · G. Kassem
German University in Cairo, Cairo, Egypt
e-mail: menna-tallah.wael@guc.edu.eg; gamal.kassem@guc.edu.eg

J. Marx Gómez et al. (eds.), *Artificial Intelligence Tools and Applications in Embedded and Mobile Systems*, Progress in IS,
https://doi.org/10.1007/978-3-031-56576-2_4

are related to different pattern types, in which each pattern type is the result of performing different data mining tasks on the data [2, 7–10]).

However, research papers that identify the pattern types related to performing different data mining tasks on event log data were hardly found. Hence, identifying different pattern types related to performing different data mining tasks on event log data is needed to understand how to apply data mining algorithms on event log data to discover business rules related to the different pattern types. Moreover, since the discovery of business rules should be based on a classification of the business rules [11], and since business rules are incorporated in patterns within event log data [3], identifying the types of patterns that can be discovered from event logs through performing different data mining tasks is also needed to classify the business rules that can be discovered from event logs based on different pattern types.

Therefore, conducting a systematic literature review is important to identify from current and previous research, different pattern types related to performing different data mining tasks on event log data, and the different analysis objectives for the discovery of business rules from event logs to guide the process analyst of the different analysis objectives that could be reached through the discovery of business rules from event logs using data mining algorithms.

This paper is structured in sections: after this introductory section, the second section is about performing the systematic literature review. In the third section, the findings of the systematic literature review are presented. Finally, the conclusion is provided in the last section.

2 Systematic Literature Review

To conduct the Systematic Literature Review (SLR), the review process proposed by Imran et al. [12] was followed which consists of three steps: review planning, review conducting, and review reporting. In the review planning step, the purpose of the review, the review questions, and the review protocol are identified. In the second step, the review is conducted and the data that is needed to answer the review questions are extracted. In the last step, the review results are reported. The first two steps are discussed in this section, while the results of conducting the review and reporting of the results will be presented in the following section.

2.1 Review Planning

As discussed earlier, the identification of the patterns within event log data related to performing different data mining tasks on event log data, and the different analysis objectives for the discovery of business rules from event logs are needed. Moreover, it was found that no systematic review was performed to identify this information previously. Therefore, the purpose of this research is to conduct an SLR to identify from current and previous literature, the pattern types within event log data related to performing different data mining tasks on event log data, which is needed to classify the discovery of rules according to the extracted pattern types. In addition to identifying the different analysis objectives for the discovery of business rules from event logs. Accordingly, the specific review questions needed to be answered are:

RQ1. What are the common pattern types related to performing different data mining tasks on event log data to discover business rules?
RQ2. What are the analysis objectives for the discovery of business rules from event logs?

After identifying the review questions, the review protocol was developed, which consists of identifying the search terms, selecting the relevant databases for searching, and formulating the inclusion and exclusion criteria. The following subsections present the details of the review protocol.

A: **Search Terms**

The review questions were used to identify the main search terms which are (patterns, rules, business rules, event log, data mining, and business process). The Boolean AND was used to connect the search terms and the Boolean OR was used to express alternative terms that can be used. Based on the search terms, the following search queries were used for searching: ("business rules" OR "rules" OR "patterns) AND ("data mining" OR "mining") AND ("business process" OR "event log").

B: **Databases to Search**

Three online scholarly and international databases were selected which are: Science Direct, IEEE Explore, and Google Scholar. The three databases were chosen as they are considered relevant databases in information technology.

C: **Inclusion and Exclusion Criteria**

The inclusion criteria were used to select papers that meet some important requirements. The inclusion criteria are:

- The research paper is written in English.
- The research paper focuses on the discovery of rules from event logs.
- The research paper uses data mining for the discovery of business rules from the event logs.

The exclusion criteria were used to remove papers that do not meet some important requirements. The exclusion criteria are:

- The Research paper focuses on the discovery of rules from sources other than event logs.
- The Research paper does not use data mining algorithms to discover the business rules.

2.2 *Conducting the SLR*

A manual search for conference proceedings and journal papers was performed from the three selected databases with the identified search queries. To conserve the ability to find as many articles as possible, the review was not limited to a specific time period. The number of articles found in the three databases are (605) articles, of which (405) articles were identified from ScienceDirect, (100) articles were identified from IEEE Explore, and (100) articles were identified from Google Scholar given the first ten pages of Google Scholar, as viewing the rest of the pages did not yield relevant articles. Then, title and abstract-based screening were performed and the duplicates were removed, this reduced the results to (100) articles. After that, full-text screening was performed according to the inclusion and exclusion criteria, which reduced the results to (40) articles. Furthermore, the quality of the articles was assessed according to the following quality criteria: the number of citations, and the degree of providing a clear methodology and a thorough discussion of the results. Accordingly, articles that had no citations were removed and articles with ambiguous methodology/results were also removed, which reduced the results to (31) articles.

The 31 relevant research papers were used to extract data that answer the review questions. Data extracted from each article were:

- The reference of the paper.
- The pattern type that was discovered from event log data.
- Explanation of the pattern discovered.
- The data mining task is related to the discovery of that pattern type.
- The objectives for the discovery of each pattern type from event log data.
- The analysis objective for the discovery of the business rules from event logs and the explanation of each analysis objective.

3 Results and Discussion

Regarding the first review question, the results of the review identified different pattern types within event log data related to performing different data mining tasks. The identified patterns within event log data resulted from the application of data mining algorithms on event log data to fulfill a data mining task. Three different patterns within event log data were commonly and repeatedly mentioned in the SLR findings which are: frequent patterns, rare patterns, and classification patterns. Depending on the data mining tasks performed on event log data, each pattern type can be discovered in the form of patterns or rules. More details regarding each pattern type within the event log data are discussed in the following subsections.

1. Frequent Patterns within Process Data

Frequent patterns within process data reflect frequently associated event log attributes [13]. This pattern type is related to performing data mining tasks such as frequent itemset mining, association rule mining, sequential pattern mining, and sequential rule mining tasks on event log data. Frequent patterns within process data can be discovered in the form of:

- Frequent association rules: which reflect frequent correlations between process attributes such as activities, resources, and time-related attributes [13, 14].
- Frequent sequential patterns/rules: which reflect correlated process attributes that are sequential or time-ordered [15–17]. Sequential patterns could be ordered subsequences of process events [16].

2. Rare Patterns within Process Data

Rare patterns within process data reflect infrequent associations of event log attributes [13]. Rare patterns are correlated attributes that occur infrequently. Rare process patterns could show an accepted behavior that should be enforced or an unacceptable behavior that should be further analyzed [18]. Rare patterns can reveal important information like risks and fraud [19–23]. Moreover, rare patterns can be detected by identifying patterns that have support values below a user-specified support threshold [20] or by detecting rules that have low support and high confidence values [24]. This pattern is related to performing data mining tasks such as rare pattern mining, rare association rule mining, and rare sequential pattern mining tasks on event logs to discover rare association/sequential rules that reflect rare process patterns.

3. Classification/Prediction Patterns within Process Data

Classification patterns within process data reflect the classification of event log attributes according to a class label attribute. Classification patterns can be discovered in the form of classification patterns or rules which show the relationship between a response attribute and a set of predictor attributes [25, 26]. Rules discovered could be used for the prediction of unseen event log class attributes [27]. This pattern is related to performing data mining tasks such as rule learning and classification pattern/rule mining tasks on event log data. Each pattern type identified from the SLR was discovered as patterns/rules that satisfy different event log analysis objectives. Examples of SLR papers that discovered each pattern type and the objectives for the discovery of each pattern type are as follows:

Objectives related to the discovery of frequent association rules

- Discovery of frequent itemsets in an event log. This analysis objective was satisfied by discovering association rules that reflect correlations between process attributes like task, temporal, or resource-related attributes [13].
- Discovery of access control policies given to resources. This analysis objective was satisfied by discovering association rules that detect frequent correlations between process tasks, process data, and the different resources [19].

- Discovery of resource allocation rules to ensure each activity is performed by the right resource. This analysis objective was satisfied by discovering association rules that reflect frequent correlations between activities and resources [28].

Objectives related to the discovery of frequent sequential patterns/rules

- Discovery of sequential relationships between process activities is needed to discover a process model [14–16]. This analysis objective was satisfied by discovering the sequential patterns within process activities.
- Discovery of profitable sequential process patterns. This analysis objective was satisfied by discovering high-utility sequential patterns within event log data [15, 16].
- Discovery of low-cost sequential process patterns that can be used to lower process costs. This analysis objective was satisfied by discovering low-cost sequential patterns within event log data [15, 16].
- Discovery of overlapping process tasks. This analysis objective was satisfied by discovering temporal patterns within event log data [29].

Analysis objectives related to the discovery of rare patterns within process data in the form of association patterns/rules or sequential patterns/rules

- Improving process discovery results [20, 22, 23]. This analysis objective was satisfied by discovering rare sequential rules within event log data, then the rules were examined to determine if they are correct or noisy patterns.
- Detecting violated data access permissions given to resources [19]. This analysis objective was satisfied by discovering rare association rules that are below the user support and confidence values and then examining these patterns.
- Detecting security violations like fraud and risks [21]. This analysis objective was satisfied by discovering rare patterns that are below the user support value and then examining these patterns.

Analysis objectives related to classification patterns within process data in the form of classification rules

- Discovery and prediction of process model constructs, in which the constructs reflect activity relations such as the parallel, exclusive choice, and causal relations. Classification rules are discovered by identifying and predicting the relations between activities through the classification of labeled classes of relations [14, 30].
- Decision point analysis that analyzes and predicts the relationship between the event log data attributes and the routing choices of process execution. Decision point analysis generates classification rules (decision rules) by classifying the event log data attributes according to a class attribute which is the different routing decisions that could be made [31–33].
- Discovery of Resource patterns that identify the required profile of resources to perform a process activity instance. Classification rules are discovered by classifying attributes like the process tasks, organizational agents, organizational

positions, and roles into performers or non-performers of a process activity instance [34].
- Discovery of classification rules that reflect the root cause of risk incidents like overtime cases [26] or analyze and explain the deviant process cases by classifying cases into deviant or normal [35–37].

Regarding the second review question, the results of the review identified a list of different analysis objectives for the discovery of business rules from event logs, the following list summarizes the different analysis objectives and explains each business rules analysis objective:

- Process model constructs analysis and prediction: To discover rules related to process model constructs like sequences, parallel, and exclusive choices [30].
- Process performance analysis: To discover rules that detect patterns within process data related to the cost, time, or quality attributes [38].
- Decision point analysis: To discover decision rules that capture the effect of data attributes on the routing of a process instance, by analyzing and predicting the relationship between the event log data attributes and the routing choices of a case [31].
- Discovery of low-cost/high-utility process patterns: To discover rules that detect profitable or low-cost patterns within process data [15, 16].
- Causal relationship discovery: To discover causal rules that detect causal relationships between event log data attributes [14].
- Root cause analysis: To discover rules that explain the reason behind a specific process phenomenon or behavior by analyzing event log data attributes [26].
- Temporal patterns discovery: To discover rules that describe temporal patterns within process data such as temporal overlapping patterns within process data [29].
- Resource performance analysis: To discover rules that reflect resource behavioral patterns within process data. Discovered rules can reflect skills, productivity, and utilization of resources [39].
- Staff assignment/Resource allocation rules discovery: To discover rules that determine the profile of a resource needed to perform a certain activity. Business rules discovered help managers allocate the right resources to the right tasks and allocate alternative resources to a specific task if a required resource is busy [28]. Moreover, the business rules discovered help managers ensure each activity is performed by the right resource [28, 34, 40].
- Organizational roles analysis: To discover rules that determine patterns within process data related to organizational roles, permissions, rights, and the interaction between different roles [41].
- Organizational structure analysis: To discover rules that determine patterns within process data that reflect the structure of the organization [41].
- Detection of violated process policies: To discover rules related to infrequent patterns within process data that might indicate fraud, violations, or deviation from the accepted behavior [19, 21, 42].

- Access control rules discovery: To discover rules that detect patterns within process data that are related to granted access permissions given to users. The discovery aims to protect against unauthorized access to process data [19].

4 Conclusion and Outlook for Future Work

This paper conducted a systematic literature review by running search queries on three databases to answer two review questions (see Sect. 2). Regarding RQ1, we concluded that three pattern types were commonly discovered from event logs as a result of applying data mining algorithms on event log data to discover business rules. The three patterns identified are the frequent patterns, the rare patterns, and the classification patterns. Regarding RQ2, we concluded that there are different analysis objectives for the discovery of business rules from event logs. The results of this SLR shed light on the different pattern types within the event log data related to the discovery of business rules, the different analysis objectives for the discovery of business rules, and the possible data mining methods that could be used to discover them in event logs. We think that these results will support the development of frameworks that aim to standardize the process of business rules discovery from event logs using data mining algorithms, the frameworks can then be used to design future process mining analysis systems.

Limitations of this research include missing some relevant references due to performing a manual search process, selecting only three databases for the search, and selecting the articles written only in the English language. Accordingly, for future work, it is recommended to perform an automated search instead of the manual search, to include more databases for the search, and to search for articles in languages other than English language.

References

1. Graml, T., Bracht, R., & Spies, M. (2007, October). Patterns of business rules to enable agile business processes. In *11th IEEE International Enterprise Distributed Object Computing Conference (EDOC 2007)* (pp. 365–365). IEEE.
2. Polpinij, J., Ghose, A., & Dam, H. K. (2015). Mining business rules from business process model repositories. *Business Process Management Journal, 21*(4), 820–836.
3. Thi, T. T. P., Helfert, M., Hossain, F., & Dinh, T. L. (2011, June). Discovering business rules from business process models. In *Proceedings of the 12th International Conference on Computer Systems and Technologies* (pp. 259–265).
4. Crerie, R., Baião, F. A., & Santoro, F. M. (2009). Discovering business rules through process mining. In *Enterprise, business-process and information systems modeling* (pp. 136–148). Springer.
5. Bolt, A., de Leoni, M., & van der Aalst, W. M. (2018). Process variant comparison: Using event logs to detect differences in behavior and business rules. *Information Systems, 74*, 53–66.

6. Salama, M., & Kassem, G. (2021). A data analytics approach for game-based physical therapy systems. In *AMCIS*.
7. Polpinij, J., Ghose, A. K., & Dam, H. K. (2010, July). Business rules discovery from process design repositories. In *In 2010 6th world congress on services* (pp. 614–620). IEEE.
8. Corea, C., & Delfmann, P. (2020). A taxonomy of business rule organizing approaches in regard to business process compliance. *Enterprise Modeling and Information Systems Architectures (EMISAJ), 15*, 4–1.
9. Bose, I., & Mahapatra, R. K. (2001). Business data mining—A machine learning perspective. *Information & Management, 39*(3), 211–225.
10. Swinnen, J., Depaire, B., Jans, M. J., & Vanhoof, K. (2011, August). A process deviation analysis–a case study. In *International conference on business process management* (pp. 87–98). Springer.
11. Van der Linden, E., Smit, K., Berkhout, M., & Zoet, M. (2019). The business rule type jungle: An explorative analysis (2019). *BLED 2019 Proceedings*, 24.
12. Imran, M., Ismail, M. A., Hamid, S., & Nasir, M. H. N. M. (2022). Complex process modeling in process mining: A systematic review. *IEEE Access, 10*, 101515–101536.
13. Djenouri, Y., Belhadi, A., & Fournier-Viger, P. (2018). Extracting useful knowledge from event logs: A frequent itemset mining approach. *Knowledge-Based Systems, 139*, 132–148.
14. Acheli, M., Grigori, D., & Weidlich, M. (2021). Discovering and analyzing contextual behavioral patterns from event logs. *IEEE Transactions on Knowledge and Data Engineering, 34*(12), 5708–5721. https://doi.org/10.1109/TKDE.2021.3077653
15. Dalmas, B., Fournier-Viger, P., & Norre, S. (2017). Twinkle: A constrained sequential rule mining algorithm for event logs. *Procedia Computer Science, 112*, 205–214.
16. Fournier-Viger, P., Li, J., Lin, J. C. W., Chi, T. T., & Kiran, R. U. (2020). Mining cost- effective patterns in event logs. *Knowledge-Based Systems, 191*, 105241.
17. Özdağoğlu, G., Öztaş, G. Z., & Çağliyangil, M. (2019). An application framework for mining online learning processes through event-logs. *Business Process Management Journal, 25*(5), 860–886.
18. Bezerra, F., & Wainer, J. (2013). Algorithms for anomaly detection of traces in logs of process aware information systems. *Information Systems, 38*(1), 33–44.
19. Abou Rida, A., Assy, N., & Gaaloul, W. (2019). Extracting attribute-based access control rules from business process event logs. In *BDCSIntell* (pp. 38–45).
20. Fani Sani, M., Zelst, S. J. V., & van der Aalst, W. M. (2018, October). Applying sequence mining for outlier detection in process mining. In *OTM confederated international conferences on the move to meaningful internet systems* (pp. 98–116). Springer.
21. Mannhardt, F., Leoni, M. D., Reijers, H. A., & van der Aalst, W. M. (2017, June). Data- driven process discovery-revealing conditional infrequent behavior from event logs. In *International conference on advanced information systems engineering* (pp. 545–560). Springer.
22. Marin-Castro, H. M., & Tello-Leal, E. (2021). Event log preprocessing for process mining: A review. *Applied Sciences, 11*(22), 10556.
23. Pourmasoumi, A., Kahani, M., & Bagheri, E. (2017). Mining variable fragments from process event logs. *Information Systems Frontiers, 19*(6), 1423–1443.
24. Romero, C., Luna, J. M., Romero, J. R., & Ventura, S. (2011). RM-tool: A framework for discovering and evaluating association rules. *Advances in Engineering Software, 42*(8), 566–576.
25. Khanbabaei, M., Sobhani, F. M., Alborzi, M., & Radfar, R. (2018). Developing an integrated framework for using data mining techniques and ontology concepts for process improvement. *Journal of Systems and Software, 137*, 78–95.
26. Suriadi, S., Ouyang, C., van der Aalst, W. M., & ter Hofstede, A. H. (2012, September). Root cause analysis with enriched process logs. In *International conference on business process management* (pp. 174–186). Springer.
27. Márquez-Chamorro, A. E., Resinas, M., & Ruiz-Cortés, A. (2017). Predictive monitoring of business processes: A survey. *IEEE Transactions on Services Computing, 11*(6), 962–977.

28. Liu, T., Cheng, Y., & Ni, Z. (2012). Mining event logs to support workflow resource allocation. *Knowledge-Based Systems, 35*, 320–331.
29. Hwang, S. Y., Wei, C. P., & Yang, W. S. (2004). Discovery of temporal patterns from process instances. *Computers in Industry, 53*(3), 345–364.
30. Măruşter, L., Weijters, A. J., Van Der Aalst, W. M., & Van Den Bosch, A. (2006). A rule-based approach for process discovery: Dealing with noise and imbalance in process logs. *Data Mining and Knowledge Discovery, 13*(1), 67–87.
31. Rozinat, A., & van der Aalst, W. M. (2006, September). Decision mining in ProM. In *In the international conference on business process management* (pp. 420–425). Springer.
32. Sarno, R., Sari, P. L. I., Ginardi, H., Sunaryono, D., & Mukhlash, I. (2013, November). Decision mining for multi choice workflow patterns. In *2013 international conference on computer, control, informatics and its applications (IC3INA)* (pp. 337–342). IEEE.
33. Sarno, R., Sari, P. L. I., Sunaryono, D., Amaliah, B., & Mukhlash, I. (2014, September). Mining decision to discover the relation of rules among decision points in a non-free choice construct. In *Proceedings of international conference on information, communication technology and system (ICTS)* (pp. 53–58). IEEE.
34. Ly, L. T., Rinderle, S., Dadam, P., & Reichert, M. (2005, September). Mining staff assignment rules from event-based data. In *International conference on business process management* (pp. 177–190). Springer.
35. Genga, L., Potena, D., Chiorrini, A., Diamantini, C., & Zannone, N. (2020). A latitudinal study on the use of sequential and concurrency patterns in deviance mining. In *Complex pattern mining* (pp. 103–119). Springer.
36. Gupta, N., Anand, K., & Sureka, A. (2015, March). Pariket: Mining business process logs for root cause analysis of anomalous incidents. In *International workshop on databases in networked information systems* (pp. 244–263). Springer.
37. Nguyen, H., Dumas, M., Rosa, M. L., Maggi, F. M., & Suriadi, S. (2014, October). Mining business process deviance: A quest for accuracy. In *OTM confederated international conferences on the move to meaningful internet systems* (pp. 436–445). Springer.
38. Van der Aalst, W. (2016). Data science in action. In *Process mining* (pp. 3–23). Springer.
39. Pika, A., Leyer, M., Wynn, M. T., Fidge, C. J., Hofstede, A. H. T., & Aalst, W. M. V. D. (2017). Mining resource profiles from event logs. *ACM Transactions on Management In- formation Systems (TMIS), 8*(1), 1–30.
40. Huang, Z., Lu, X., & Duan, H. (2011). Mining association rules to support resource allocation in business process management. *Expert Systems with Applications, 38*(8), 9483–9490.
41. Zhao, W., & Zhao, X. (2014). Process mining from the organizational perspective. In *Foundations of intelligent systems* (pp. 701–708). Springer.
42. Van der Aalst, W. M., & de Medeiros, A. K. A. (2005). Process mining and security: Detecting anomalous process executions and checking process conformance. *Electronic Notes in Theoretical Computer Science, 121*, 3–21.

Contextual Multi-View Graph Community Detection Using Graph Neural Networks

Chiheb Edine Zoghlemi and Abdelkerim Rezgui

1 Introduction

Clustering is one of the most important topics in the domain of data analysis, exploration, and visualization. It is an unsupervised technique for the classification of samples that aims to group similar samples into one group called a cluster [1]. Roughly speaking, clustering consists of discovering natural groups of similar traits in a dataset.

An interesting and important variant of data clustering is graph clustering. One of the most significant applications of graph clustering is community detection. This application consists of extracting dense and similar communities (clusters) from the network (graph). The vertices in each community are more densely connected internally than the rest of the network. In the last decade graphs exhibited increased usage, as the graph properties capture different complex data and relations. This demand led to a growing attention from the literature on graph analysis, exploration, and clustering techniques. Graphs can vary in size (number of edges and nodes), features (number of features by edge or node), or structure (multiple edge types, multiple node types or both). In this work, we will focus on the multi-view graphs. Multi-view graph is a network, composed of the same vertices but with different edge types, every view is composed of an edge type that represents a subset of relations between the vertices.

Applying a community detection technique on a multi-view graph is a complex task itself, especially when the importance of certain view relations is more crucial than the remaining views. The importance of these views is evaluated based on the business needs and demands. In the industry, these needs and demands are

C. E. Zoghlemi (✉) · A. Rezgui
BI4YOU, CUN A5.2 Golden Towers, Tunis, Tunisia
e-mail: c.zoghlami@biforyou.com; a.rezgui@biforyou.com

J. Marx Gómez et al. (eds.), *Artificial Intelligence Tools and Applications in Embedded and Mobile Systems*, Progress in IS,
https://doi.org/10.1007/978-3-031-56576-2_5

rapidly changing, as these demands are linked to a certain business context. The fact that certain views may be important under certain contexts should not neglect the importance and information that resides within the remaining views. To tackle the community detection problem with consideration of a certain context, in this paper we introduce a novel approach called Contextual Graph Clustering (CGC). This approach enables applying the community detection problem under a dynamic and rapidly changing context.

CGC takes into consideration every view, importance, and information in the community detection process to determine the communities' preferences.

In this work, we begin by providing basic definitions and related works in Sect. 2. In Sects. 3 and 4 the novel approach will be presented. The experiments and the evaluation are discussed in Sect. 5. The conclusion and perspectives are summarized in Sect. 6.

2 Preliminaries

The CGC approach is built using two main components, embedding and community detection. For the embedding part, we implemented a graph representational learning technique. This allows the relational knowledge of interacting entities to be extracted and accessed efficiently. "Various graph embedding techniques have been developed to convert the raw graph data into a high-dimensional vector while preserving intrinsic graph properties. This process is also known as graph representation learning. With a learned graph representation, one can adopt machine-learning tools to perform downstream tasks conveniently" [2]. Several methods to tackle graph representation learning can be grouped under three categories: random-walk, graph factorization, and graph neural networks.

DeepWalk and node2vec [3] are random-walk-based techniques. They consist of extracting a large number of paths by starting from random initial nodes. These paths indicate the context of connected vertices. The randomness of walks gives the ability to explore the graph and capture both the global and the local structural information by walking through neighboring vertices. After building the paths, probability models such as skip-gram [4] and bag-of-words [5] can be performed on these randomly sampled paths to learn the node representation.

Graph Laplacian eigenmaps [6], Node proximity matrix factorization [7] are techniques based on graph factorization, more specifically on factoring the adjacency matrix of a graph. By applying the factorization process the graph structure is preserved and also the dimension of the graph is reduced. Inspired by the success of Recurrent neural networks (RNNs) and Convolutional Neural Networks (CNNs) in tackling the dimension reduction problem. In order generalize these models to graphs Graph Neural Networks (GNN) was implemented. Graph Auto Encoder (GAE) [8], Variational graph auto-encoders (VGAE) [8], GraphSAGE [9], and Structural Deep Network Embedding (SDNE) [10] are GNN's variants that aim to tackle the graph representation problem.

In addition to the first component embedding, the second one, community detection aims to detect dense and similar clusters/groups in a network. The nodes in each cluster are more densely connected internally than the rest of the network. In fact, community detection is a fundamental problem in social network analysis. It consists of dividing social actors (modeled as nodes in a social graph) with certain social connections (modeled as edges in the social graph) into densely and highly related separated groups [11].

Tackling the community detection problem is not only restricted to the stated methods. Communities also can be extracted using a clustering algorithm such as K-means [12]. In fact, K-means is an iterative technique that tries to split the dataset into K pre-defined distinct communities where each data point belongs to only one group. It tries to make the intra-cluster data points as similar as possible while also keeping the communities as different as possible. It assigns data points to a cluster such that the sum of the squared distance between the data points and the cluster's centroid is at the minimum.

3 Community Detection Methods

The literature tackled the community detection problem with graph neural networks on numerous occasions. Due to the variety of information, the different network topologies, and the different graph characteristics, different types of approaches have been proposed. Some approaches combine the power of Convolution Neural Networks (CNN) and Auto-encoder (AE) to tackle the community detection problem. AE + CNN-based model [13] is an auto-encoder and CNN-based deep community detection method for social networks. To extract spatial proximity matrices and generate reconstructed adjacency matrices, this work proposes a new matrix reconstruction method based on a novel cyberspace structure reconstruction. The matrix adds spatial proximity to the traditional adjacency matrix, obtains clear subspace characteristics, and makes convolutional operations to extract graph spatial localization. The AE + CNN-based model is constructed to learn spatial eigenvectors that automatically extract the spatial features of graphs. As the last step, the author used records from the reconstructed adjacency matrix and applied the K-means algorithm for detecting communities in a social network.

Some approaches use a GCN-based model for community detection, from which we find the Graph Unsupervised Community Detection in attribute networks (GUCD) approach [14]. This approach incorporated the network modeling method of MRFasGCN [15] in the Auto-encoder and introduced a different neural network architecture suitable for learning network communities and node semantics at the same time. They introduced a local enhancement to the latent communities by constructing an aggregated graph with the information from both topology and attributes, and then made each pair of connected nodes in this graph have similar community distributions.

Multi-View Attribute Graph Convolution Networks for Clustering (MAGCN) approach [16] contains two principal parts, The multi-view attribute graph con-

Table 1 Sample of the existing community detection methods

Method	GNN	Node attributes	Edge attributes	Multi- Layer	Dynamic context
AE + CNN-based model	X	–	–	–	–
GUCD	X	X	–	–	–
MAGCN	X	X	X	X	–
GAER	X	X	–	–	–
CGC	X	X	X	X	X

volutional Encoders part and the clustering part which is eventually conducted on the ideal embedding extracted from the first part. MAGCN is designed with dual encoders that reconstruct the extracted features in high dimensions and integrate the low-dimension consistent information. Multi-view attribute graph auto-encoder and consistent embedding encoder network successively reduce the noise and the difference among different views, and finally get the ideal description space of multi-view attribute graph for clustering.

In addition, some approaches use graph auto-encoders for community detection, GAER [17] is a novel community detection approach based on graph auto-encoder reconstruction. Using the modularity optimization and neighborhood Laplacian smooth sampling, by jointly using node modularity features and neighborhood information, they improved the accuracy of GAER in community detection.

We summarize the following methods in Table 1 in terms of five desirable properties, the usage of graph neural networks, the usage of node features, the usage of edge feature, their capability to handle multi-view graph, and finally the existence of the context aspect.

4 Contextual Graph Clustering

Through this section, we will introduce a novel approach, Contextual Graph Clustering (CGC). This approach is inspired by the divide-and-conquer method (DandC). It is a strategy of solving a large problem by breaking it into smaller subproblems, solve the subproblems recursively, and then combine these solutions to create a solution to the original problem [18]. In our case, we divided the network based on the views, then feed every view separately to an embedding block. The embedding block will transform the view features into a lower space representation, which will be combined later alongside the context coefficients in the fusion and context block. As a last step, the approach will extract the communities using a clustering block.

CGC applies the context directly on every view embedding by multiplying the embedding with the associated context coefficient. The context coefficient is a numerical representation of the importance of a view in comparison to the current context. Its values can vary from -1 to 1. The closer the value to 1 the more

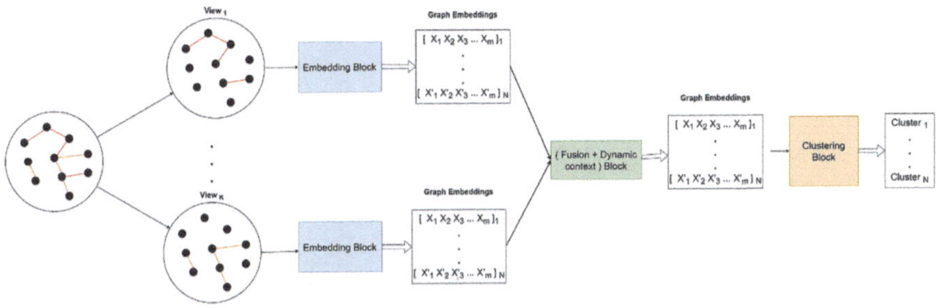

Fig. 1 Contextual graph clustering components

important it is and vice versa, whereby the value 0 is neutral. The context is applied using the following equation:

$$E'(v) + E(v) \times C(v)$$

E(v) is the view embeddings, C(v) represents the view context coefficient, and E'(v) is the new calculated embedding with context. In the next step, we calculate the graph embedding by applying the average function to the view's embeddings. In the last step, we feed the average embedding to a clustering algorithm (K-means) to detect the communities. In Fig. 1, we show the recapitulation on the approach's steps.

As we stated, this approach consists of three blocks, embedding blocks responsible for transforming the graph view into lower dimension representation. Inspired by the success realized with graph auto-encoders in graph representational learning, we used GAE [2] as the core component of this block. The fusion block integrates the view context coefficients in the embedding, then it aggregates the embedding using the average function. Figure 2 is a visualization of the steps on the fusion and context block.

The clustering block, presented in Fig. 3, applies a clustering algorithm on the embedding to detect the communities. Due to its simplicity and proven performance in several application domains, the clustering algorithm used in this block is the k-means algorithm [19].

To determine the K value in the algorithm, we evaluated a range of K values using the silhouette coefficient metric, the k value that scored the best result will represent the number of communities. Silhouette coefficient evaluates the results of communities in terms of how well samples are clustered within the same community and with comparison to other communities. The silhouette score value ranges from -1 to 1, a value closer to 1 means that the cluster is dense and well-separated than other clusters, a value closer to 0 represents overlapping clusters with samples very close to the decision boundary of the neighboring clusters, and negative score indicates that the samples might have got assigned to the wrong clusters. The silhouette score S, is calculated using the following equation:

$$S = \frac{(b - a)}{\max(a, b)}$$

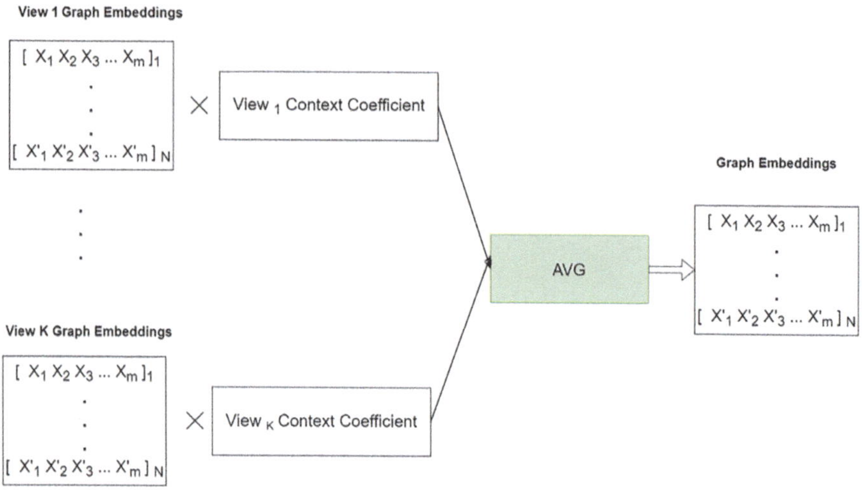

Fig. 2 CGC fusion and context block

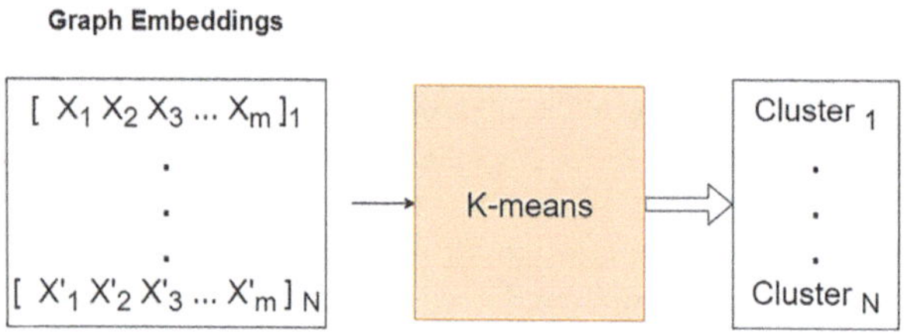

Fig. 3 CGC clustering block

a represent the mean distance between the observation and all other data points in the same community. This distance can also be called the mean intra-cluster distance. Whereby ***b*** indicates the mean distance between the observation and all other data points of the next nearest cluster. This distance can also be called the nearest-cluster distance.

5 Experiments

For experiment, and due to the lack of multi-view graph datasets and their corresponding context, the dataset used in this work was provided by BI4YOU, the dataset captures information about individuals under the social security domain. The

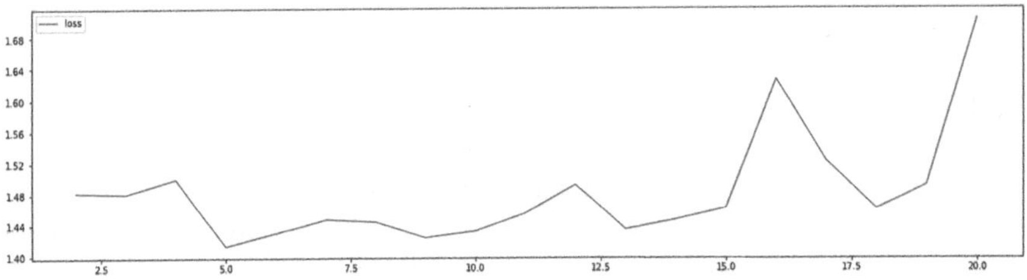

Fig. 4 Average embeddings loss score

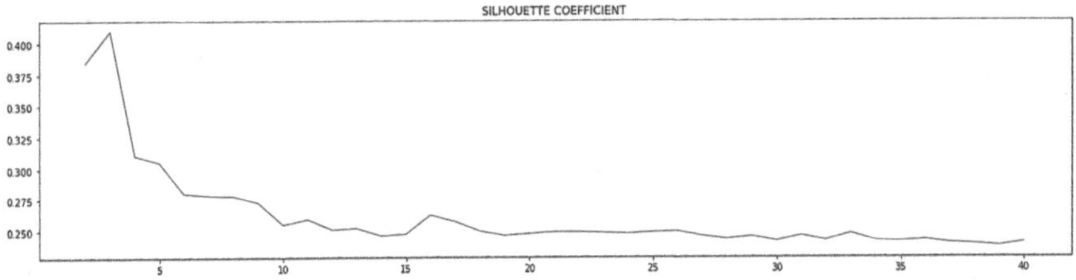

Fig. 5 Number of communities in CGC

dataset is composed of 14,254 objects and 5 views.[1] In this section, we introduce the experiments conducted within every block.

5.1 Embedding Block

The embedding size in the embedding block is determined by calculating the average loss obtained from all views within a range of dimensions. As presented in Fig. 4, the 5-dimension representation attained the lowest loss score within the current dataset.

5.2 Clustering Block

The number of communities is determined by conducting an evaluation using a range of values. The most adequate value is the value that achieved the best silhouette score. Fig. 5 presents the evaluation results using the current dataset. We notice that the best score is attained with three communities with a silhouette score equal to 0.415.

[1] The dataset is provided under a sensitive domain and for data protection reasons the data is anonymous and no data description were provided.

5.3 *Fusion and Context Block*

To illustrate the process of merging the context with the embeddings, we created the embedding using a 3-demension representation, and focused on only one node.

After applying the following context coefficients in Table 2, we concluded the following result represented in Fig. 6. In this figure, we observe that the **contextual average embedding** is closer to the **View 1** embedding than the **average embedding**.

Table 2 Sample context coefficients

View	View 1	View 2	View 3	View 4	View 5
Coefficient	1	0	0	0	0

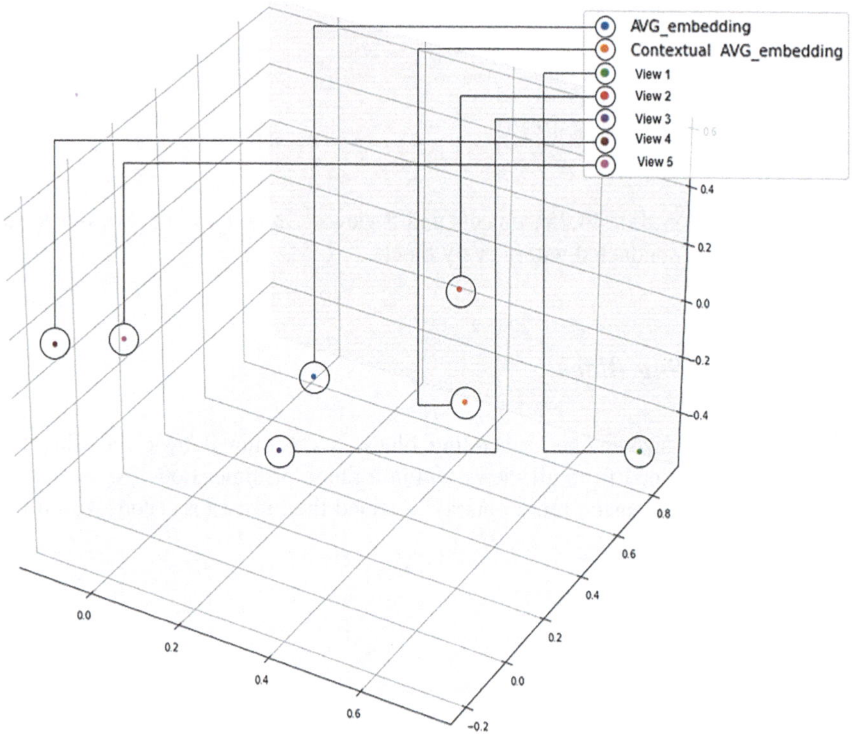

Fig. 6 3-Demension contextual embeddings

6 Conclusion

In this work, we introduced a novel approach CGC, which integrates the dynamic context to the graph clustering problem, within multi-view graphs. We also presented the view context coefficients, a numerical representation that measures the importance of a certain view to the existing context.

Through an empirical study conducted on the proposed approaches, we determined that CGC, determine communities and manipulate their preferences based on different contexts. We hope that this work will lead to further advancements and research in unsupervised learning using GNNs as well as contextual graph clustering, allowing further advances in graph learning.

Acknowledgment This work is conducted and funded by BI4YOU, a BI and AI-focused company, to tackle a real-world business. We would like to express our gratitude to the company for providing this opportunity.

References

1. Alelyani, S., Tang, J., & Liu, H. (2018). Feature selection for clustering: A review. *Data Clustering*, 29–60.
2. Chen, F., Wang, Y.-C., Wang, B., & Kuo, C.-C. J. (2020). Graph representation learning: A survey. *APSIPA Transactions on Signal and Information Processing, 9*(1). https://doi.org/10.1017/ATSIP.2020.13
3. Grover, A., & Leskovec, J. (2016, July 03). node2vec: Scalable feature learning for networks. arXiv. Retrieved June 10, 2022, from [Online]. http://arxiv.org/abs/1607.00653
4. Guthrie, D., Allison, B., Liu, W., Guthrie, L., & Wilks, Y., A closer look at skip-gram modelling, p. 4.
5. Understanding bag-of-words model: A statistical framework | SpringerLink. Retrieved August 23, 2022, from https://link.springer.com/article/10.1007/s13042-010-0001-0
6. Belkin, M., & Niyogi, P. (2003, June). Laplacian Eigenmaps for dimensionality reduction and data representation. *Neural Computation, 15*(6), 1373–1396. https://doi.org/10.1162/089976603321780317
7. Singh, A. P., & Gordon, G. J., Relational learning via collective matrix factorization, p. 9.
8. Kipf, T. N., & Welling, M. (2016, November 21). Variational graph auto-encoders. arXiv. [Online]. Retrieved June 10, 2022, from http://arxiv.org/abs/1611.07308
9. Hamilton, W. L., Ying, R., & Leskovec, J. (2018, September). Inductive representation learning on large graphs. arXiv, arXiv:1706.02216. https://doi.org/10.48550/arXiv.1706.02216.
10. Wang, D., Cui, P., & Zhu, W. (2016, August). Structural Deep Network Embedding. In *Proceedings of the 22nd ACM SIGKDD International Conference on Knowledge Discovery and Data Mining, San Francisco California USA* (pp. 1225–1234). https://doi.org/10.1145/2939672.2939753
11. Chunaev, P. (2020, August). Community detection in node-attributed social networks: A survey. *Computer Science Review, 37*, 100286. https://doi.org/10.1016/j.cosrev.2020.100286
12. Li, Y., Jia, C., & Yu, J. (2015). A parameter-free community detection method based on centrality and dispersion of nodes in complex networks. *Physica A: Statistical Mechanics and its Applications, 438*, 321–334.

13. Wu, L., Zhang, Q., Chen, C.-H., Guo, K., & Wang, D. (2020). Deep learning techniques for community detection in social networks. *IEEE Access, 8*, 96016–96026. https://doi.org/10.1109/ACCESS.2020.2996001
14. He, D., et al. (July 2020). Community-centric graph convolutional network for unsupervised community detection. In *Proceedings of the Twenty-Ninth International Joint Conference on Artificial Intelligence, Yokohama, Japan* (pp. 3515–3521). https://doi.org/10.24963/ijcai.2020/486
15. Jin, D., Liu, Z., Li, W., He, D., & Zhang, W. (2019, July). Graph convolutional networks meet Markov random fields: Semi-supervised community detection in attribute networks. In *Proceedings of the AAAI conference on artificial intelligence* (Vol. 33, pp. 152–159). https://doi.org/10.1609/aaai.v33i01.3301152
16. Cheng, J., Wang, Q., Tao, Z., Xie, D., & Gao, Q. (July 2020). Multi-View attribute graph convolution networks for clustering. In *Proceedings of the Twenty-Ninth International Joint Conference on Artificial Intelligence, Yokohama, Japan* (pp. 2973–2979). https://doi.org/10.24963/ijcai.2020/411
17. Qiu, C., Huang, Z., Xu, W., & Li, H. (March 2022). Fast community detection based on graph autoencoder reconstruction. In *2022 7th International Conference on Big Data Analytics (ICBDA)* (pp. 265–271). https://doi.org/10.1109/ICBDA55095.2022.9760357
18. Smith, D. R. (1985, January). The design of divide and conquer algorithms. *Science of Computer Programming, 5*, 37–58. https://doi.org/10.1016/0167-6423(85)90003-6
19. Ahmed, M., Seraj, R., & Islam, S. M. S. (2020). The k-means algorithm: A comprehensive survey and performance evaluation. *Electronics, 9*(8), 1295.

Feature Selection Approach to Improve Malaria Prediction Model's Performance for High- and Low-Endemic Areas of Tanzania

Martina Mariki, Neema Mduma, and Elizabeth Mkoba

1 Introduction

1.1 Malaria Burden and Diagnosis

Malaria is caused by the plasmodium parasite transmitted by an infected female anopheles mosquito bite. According to the World Health Organization (WHO), malaria remains a significant cause of death, especially in sub-Saharan Africa, with about 228 million malarial cases worldwide and an estimated 405,000 deaths in 2018 [1, 2]. Chacky et al.'s [3] showed that malaria prevalence in Tanzania was varied, with as high as 28% in the Western zone and as low as 1% in the Northern zone [3]. Ninety-three percent (93%) of mainland Tanzania's population resides in malaria-endemic areas. In 2015, there were estimated to be 7.3 million clinical and confirmed cases of malaria reported in the country [4]. The WHO has issued a guideline on the diagnosis and treatment of malaria. This guideline has to be followed by health professionals in managing malaria cases [5, 6]. The 2014 Tanzania Mainland's National Guidelines for the Diagnosis and Treatment of Malaria aligns with this guideline [4]. For malaria diagnosis, both WHO and Tanzania guidelines advocate parasitological confirmation of suspected malaria cases for people of all ages with fever, headache, joint pains, malaise, vomiting, diarrhea, body aches, body weakness, poor appetite, pallor, enlarged spleen, as a diagnostic criterion. But they also stipulate that in settings where the parasitological diagnosis is not possible, the decision to provide antimalarial treatment must be

M. Mariki (✉) · N. Mduma · E. Mkoba
Nelson Mandela African Institution of Science and Technology, Arusha, Tanzania
e-mail: marikim@nm-aist.ac.tz

J. Marx Gómez et al. (eds.), *Artificial Intelligence Tools and Applications in Embedded and Mobile Systems*, Progress in IS,
https://doi.org/10.1007/978-3-031-56576-2_6

based on the probability that the illness is malaria without the guideline on how this prediction should be made [5, 7].

In the treatment of malaria, it is advised that antimalarial medications should be administered after a parasitological confirmation of the disease is done. However, drug dispensing shops that are only permitted to sell non-prescription medications frequently distribute prescription-only treatments [8–10, 4]. This resulted in drug resistance and shortage [11, 12]. The Tanzanian government has various initiatives to reduce dispensing and usage of antimalarial drugs by people who may not have malaria. Other diseases with similar symptoms are more prevalent in low-endemic areas than malaria. Poor access to primary healthcare and errors in diagnosis represent a significant challenge to global healthcare systems. One initiative is the "not every fever is Malaria" campaign, which aims to educate people that not every fever episode is a malaria case since malaria shares similar symptoms with other febrile diseases such as dengue fever, typhoid fever, common cold, respiratory tract infection, dyspepsia, and pneumonia [13]. Since one method alone cannot eradicate the problem at hand, the need to enhance these efforts by providing tools to help predict malaria cases is most important.

Feature Selection

Machine learning has been successfully applied in predicting diseases such as cancer, diabetes, typhoid, respiratory tract infection, and even COVID-19 [14]. The study by [15] highlighted machine learning and deep learning approaches successfully applied to enhance medical care worldwide, addressing problems with diagnosis and disease treatment. This proves the potential of using these algorithms in the diagnosis of diseases. Identifying features (variables) associated with malaria diagnosis and treatment is vital for successful malaria prediction. Feature selection is an efficient data pre-processing technique in data mining to reduce data dimensionality [16]. Identifying the most important risk factors related to the disease in medical diagnosis is essential. Appropriate feature identification helps remove unnecessary, redundant attributes from the disease dataset, giving quick and better prediction results [15, 17]. This study aimed to identify significant features for malaria diagnosis both in low and high-endemic areas in Tanzania.

To begin with, the malaria diagnosis features and their importance were identified for both high and low-endemic areas. Then the performance of the machine learning algorithms was computed on the dataset with only important features to test the effectiveness of these features. Lastly, the features were verified by the medical practitioners.

2 Materials and Methods

2.1 Study Area

Data were collected from four hospitals in two regions in Tanzania: Morogoro and Kilimanjaro. The four health facilities are Mawenzi Regional Hospital, Majengo Health Centre in Kilimanjaro and Morogoro Regional Hospital, and Mzumbe Health Centre in Morogoro. The dataset represents the patients who live in the areas with low malaria transmission, represented by the Kilimanjaro region, and those who live in the areas with high malaria transmission, represented by the Morogoro region. The choice of these regions was based on the prevalence of malaria, where Morogoro region represents regions with a high prevalence of (15.0%), and Kilimanjaro represents regions with a low prevalence of (1.0%) malaria.

The Method Used and Participants

The malaria patient's records extraction form was designed based on the summary of the Ministry of Health (MoH) patient's file and the information collected when the patient visits the health facility. The records were retrieved from the patient's files who have been treated for malaria from 2015 to 2019. The aim was to identify the past state of clinical malaria diagnosis in the local health facilities and understand the standard practice in malaria diagnosis and treatment. The vital information collected was: (i) the patient's demographic information, (ii) the symptoms presented by the patient when consulting a doctor, (iii) the tests taken and results, (iv) diagnosis based on the laboratory results, and (v) the treatment provided. Trained nurses administered data collection, and all participants provided written informed consent.

Ethical Clearance

The study was approved by the National Institute for Medical Research (NIMR/HQ/R.8.c/Vol.I/1352). Before the malaria patients' records were collected and participants were recruited for the survey, permission to conduct the research was sought and granted by the medical officers in charge at the Regional, District, and health facility levels. Informed consent was obtained from all the patients (or accompanying parents/guardians of minors) who willingly signed the consent form after they were provided with information about the study's objectives. The study was of no greater than minimal risk and had no direct impact on patients' rights, welfare, or clinical care. Measures implemented to minimize the risk of confidentiality breaches during the study include anonymizing data records and keeping data secured, and accessible only to authorized persons.

Dataset Description

The dataset can be referred to as a collection of data containing many separate pieces of data. Still, it can train an algorithm to find predictable patterns inside the whole dataset [18]. The dataset corresponds to a single database table or a statistical data matrix. Every table column represents a particular variable, and each row corresponds to a given member of the dataset in question. For example, in the Malaria dataset, the variables collected included the date of visiting the facility, age, sex, residence area, symptoms observed, type of test taken, test results, diagnosis, and treatment provided, as shown in Table 1. These variables were collected based on the items recorded in the patients' files by the doctors when the patients visited the health facilities. The residence area was selected to observe if patients from a particular area have higher infection rate than the other location. The visit date was collected to observe seasonal malaria, and the symptoms were gathered to see if malaria diagnosis.

2.2 *Dataset Pre-processing*

Data cleaning, transformation, and reduction were performed on the dataset. For data cleaning, missing values were handled by removing the tuples with the missing values. Next, concept hierarchy generalization [19, 20] was used to transform the patient's residence area variable by grouping the residence area to the hospital the patient attended. Finally, feature selection as a one-dimensionality reduction technique [21, 22] was applied to reduce the number of subset attributes insignificant to the target variable. In this case, the target variable is malaria diagnosis, which can be positive or negative.

2.3 *Feature Selection*

This study used a model-based feature selection method that uses supervised machine learning algorithms to judge the importance of each feature in the dataset and select the most important features. Feature selection is one of the important processes for machine learning because including irrelevant features affects the classification performance of the machine learning model. In model-based feature selection, there are two approaches: using feature importance and selecting from the model whose main aim is to choose the most significant features [23]. Random Forest algorithm (RF) was used as a feature selection algorithm. It selected important features from the malaria diagnosis dataset because it can also give its interpretation of feature importance, which can be plotted and used for selecting

Table 1 Malaria diagnosis dataset features description low-endemic area (Kilimanjaro) and high-endemic area (Morogoro)

S. No.	Feature name	Data type	Description	Domain of values
1	Residence Area	Categorical	1 = MajengoHC, 2 = MorogoroRH, 3 = MawenziRH, 4 = MzumbeHC	1, 2, 3, 4
2	Visit Date	Categorical	Date in Months	1,2,3,4,5,6,7,8,9,10,11,12
3	Age	Categorical	Age in Years	> 5 ages <95
4	Sex	Categorical	Male = 1, Female = 0	1, 0
5	Fever	Integer	Yes = 1, No = 0	1, 0
6	Sweating	Integer	Yes = 1, No = 0	1, 0
7	Fatigue	Integer	Yes = 1, No = 0	1, 0
8	Headache	Integer	Yes = 1, No = 0	1, 0
9	Shaking & Chills	Integer	Yes = 1, No = 0	1, 0
10	Muscle Pain	Integer	Yes = 1, No = 0	1, 0
11	Joint Pain	Integer	Yes = 1, No = 0	1, 0
12	General Body Malaise	Integer	Yes = 1, No = 0	1, 0
13	Chest Pain	Integer	Yes = 1, No = 0	1, 0
14	Abdominal Pain	Integer	Yes = 1, No = 0	1, 0
15	Nausea	Integer	Yes = 1, No = 0	1, 0
16	Vomiting	Integer	Yes = 1, No = 0	1, 0
17	Coughing	Integer	Yes = 1, No = 0	1, 0
18	Dizziness	Integer	Yes = 1, No = 0	1, 0
19	Confusion	Integer	Yes = 1, No = 0	1, 0
20	Backache	Integer	Yes = 1, No = 0	1, 0
21	Restless	Integer	Yes = 1, No = 0	1, 0
22	Flue	Integer	Yes = 1, No = 0	1, 0
23	Problem breathing	Integer	Yes = 1, No = 0	1, 0
24	Anemia	Integer	Yes = 1, No = 0	1, 0
25	Yellow skin	Integer	Yes = 1, No = 0	1, 0
26	Bloody stool	Integer	Yes = 1, No = 0	1, 0
27	Appetite loss	Integer	Yes = 1, No = 0	1, 0
28	Conversion	Integer	Yes = 1, No = 0	1, 0
29	Dehydration	Integer	Yes = 1, No = 0	1, 0
30	Pale	Integer	Yes = 1, No = 0	1, 0
21	Running Nose	Integer	Yes = 1, No = 0	1, 0
32	Blurred vision	Integer	Yes = 1, No = 0	1, 0
33	Pain in urination	Integer	Yes = 1, No = 0	1, 0
34	Palpation	Integer	Yes = 1, No = 0	1, 0
35	Diarrhea	Integer	Yes = 1, No = 0	1, 0
36	Diagnosis	Categorical	Positive = 1, Negative = 0	1,0

the most informative set of features in the dataset. Feature Importance refers to techniques that calculate a score for all the input features for a given model the scores simply represent the "importance" of each feature. A higher score means that the specific feature will have a larger effect on the model that is being used to predict a certain variable.

Random forest was selected for feature selection because was robust and had high performance. This classifier is considered to be a robust model based on its ability to do intensive search of features that can maximize prediction accuracy. Apart from that RF provides a good predictive performance, low overfitting, and easy interpretability [24–26]. The interpretability is given by the fact that it is straightforward to derive the importance of each variable on the tree decision. In other words, it is easy to compute how much each variable contributes to the decision. To minimize the complexity and improve the model's performance, the top 10 important features were selected for the regional datasets and 15 important features for the combined malaria dataset. The evaluation criteria applied is if the accuracy of the model trained using the dataset with the important features is higher than the full features dataset. The selected important features will be considered significant for classifying malaria and used for the development of the malaria prediction model. Three sets of features were generated from the malaria diagnosis dataset. The first feature set was derived by applying the feature selection techniques to a low-endemic dataset (Kilimanjaro Region) area) patients, the second from a dataset of only Morogoro (high-endemic area) patients, and the last from Morogoro and Kilimanjaro (combined areas) patients.

2.4 Evaluation of Selected Features

The selected features were evaluated by comparing the accuracy of the machine learning model trained using a dataset with full features (complete set) with the models developed using the three selected feature sets (Kilimanjaro, Morogoro, and combined feature sets). Six different machine learning models were developed with each feature set, and the average accuracy was computed and then used for comparison. The six classifiers chosen were: K-Nearest Neighbour (KNN), Support Vector Machine (SVM), Naïve Bayes (NB), Logistic Regression (LR), Decision Tree (DT), and Random Forest (RF) classifiers. These were chosen due to their popularity in disease diagnosis [14, 27, 28]. Apart from that, important features were validated by the health workers from the study health facilities.

3 Results

3.1 Important Feature Selection Results

Headache, age, vomiting, visiting date, fever, general body malaise, joint pain, coughing, abdominal pain, and sex in the corresponding hierarchy as depicted in Fig. 1 where the most important features in the low-endemic areas. For the high-endemic dataset, the most important features are led by the patient's age of the patients, followed by fever, abdominal pain, visit date, dizziness, vomiting, headache, sex of the patient, general body malaise, and confusion, as shown in Fig. 2. From combined dataset, the most important features are residence area of a patient, fever, age of the patient, general body malaise, visit date, headache, abdominal pain, backache, chest pain, sex of a patient, vomiting, confusion, dizziness, coughing, and joint pain as shown in Fig. 3.

3.2 Categorical Features Correlation

From the important features selected by tree-based methods, categorical features were treated numerically after encoding them. The important features selected are the patient's residence area, visit date, sex, and age. Then, the significance of each

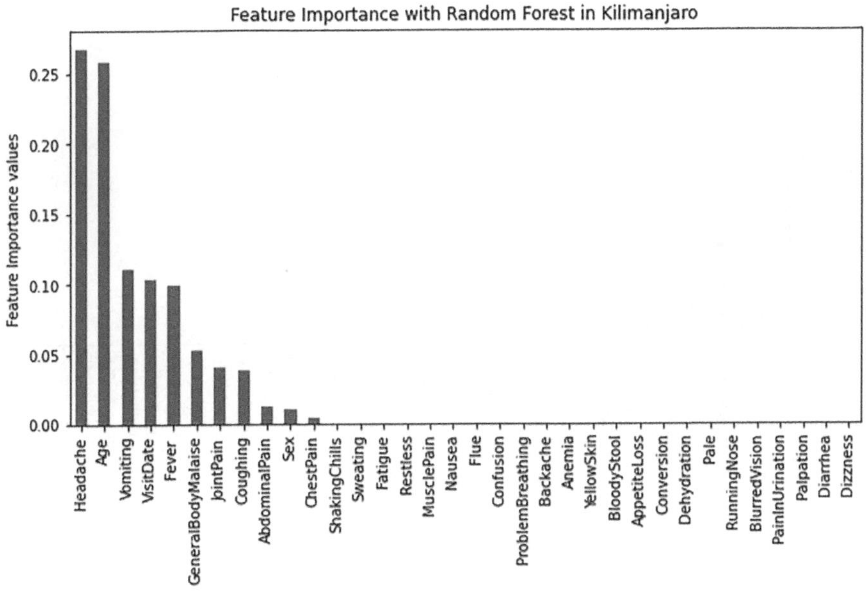

Fig. 1 Important features for low-endemic area dataset

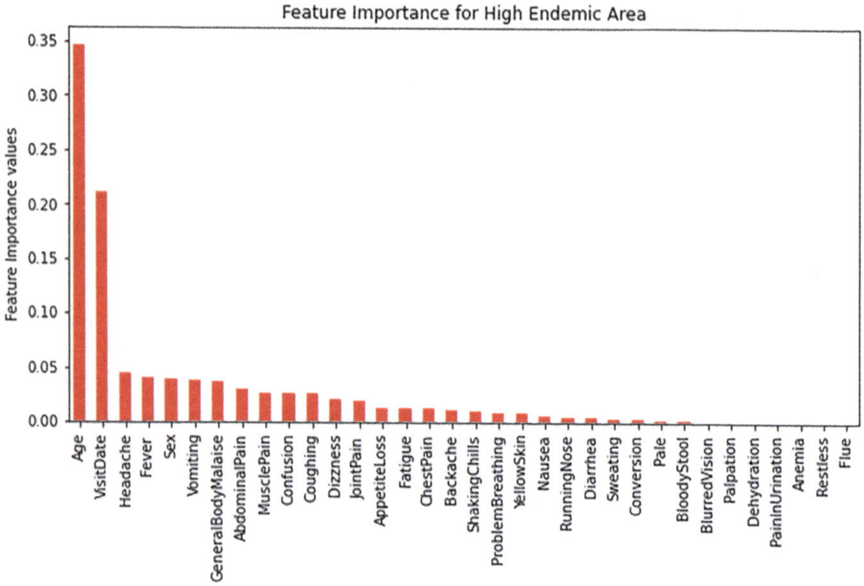

Fig. 2 Important features for high-endemic area dataset

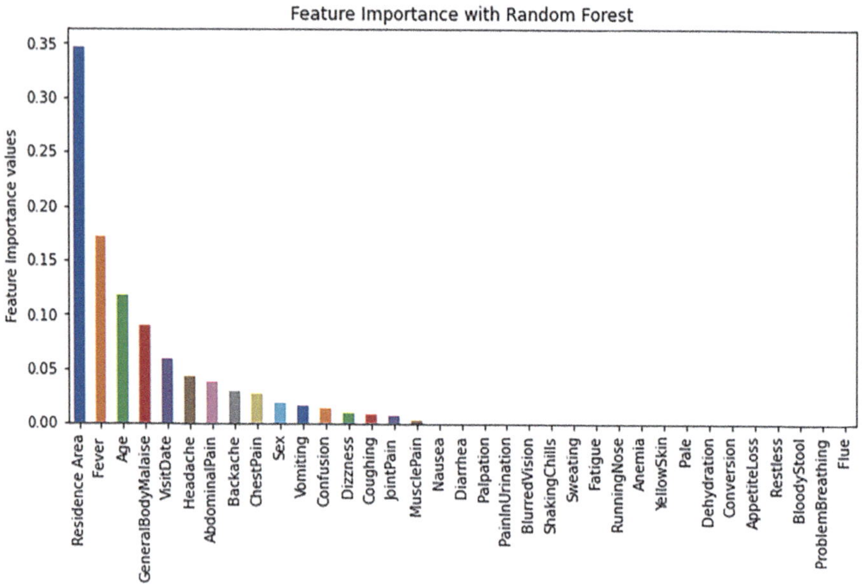

Fig. 3 Important features for combined areas dataset

Table 2 Regional-based important features in malaria diagnosis

Ranking	Full dataset	Low-endemic dataset	High-endemic dataset
1	Residence Area	Headache	Age
2	Fever	Age	Fever
3	Age	Vomiting	Abdominal Pain
4	General Body Malaise	Visit Date	Visit Date
5	Visit Date	Fever	Dizziness
6	Headache	General Body Malaise	Vomiting
7	Abdominal Pain	Joint Pain	Headache
8	Backache	Coughing	Sex
9	Chest Pain	Abdominal Pain	General Body Malaise
10	Sex	Sex	Confusion
11	Vomiting		
12	Confusion		
13	Dizziness		
14	Coughing		
15	Joint Pain		

feature and the subset of these features to the target were computed. While the sex of the patient shows no importance in diagnosing malaria in all the datasets, the residence area of the patients from the two regions showed a high significance in diagnosing malaria. Apart from that, the visit date is significant in diagnosing malaria. Therefore, the visit date variable was included in the dataset to identify if the time of the visit has any significance in the diagnosis of malaria [27, 29–32]. Basically, to confirm if there is seasonal malaria. For the Kilimanjaro region, January, February, May, and August are more significant, while Morogoro, April, July, and October are more significant than other months. Furthermore, for the combined malaria dataset, January, April, May, August, and October are significant to malaria diagnosis. Ages 12 and 55 years show significance in the malaria dataset, while ages 2, 15, and 55 years are more significant in the Morogoro and Kilimanjaro regions. Even though age has no significance in the combined dataset, other studies proved that malaria's risk increases with age (Table 2).

3.3 Important Features Evaluation

Classification Model Performance with Important Features

Random Forest and Decision Tree had the best accuracy prediction of 85% and 80%, respectively, followed by K-Nearest Neighbour with a prediction accuracy of 78%, as shown in Fig. 4. As for the low-endemic area dataset with important features, Random Forest, Decision Tree, and K-Nearest Neighbour performed well with prediction accuracy of 95%, 90%, and 82%, respectively, as shown in Fig. 5.

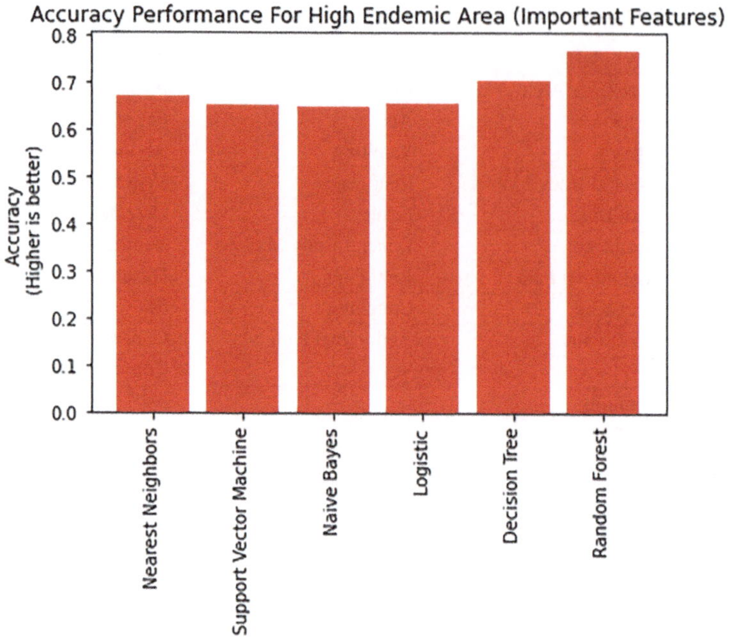

Fig. 4 Accuracy performance for important features in high-endemic dataset

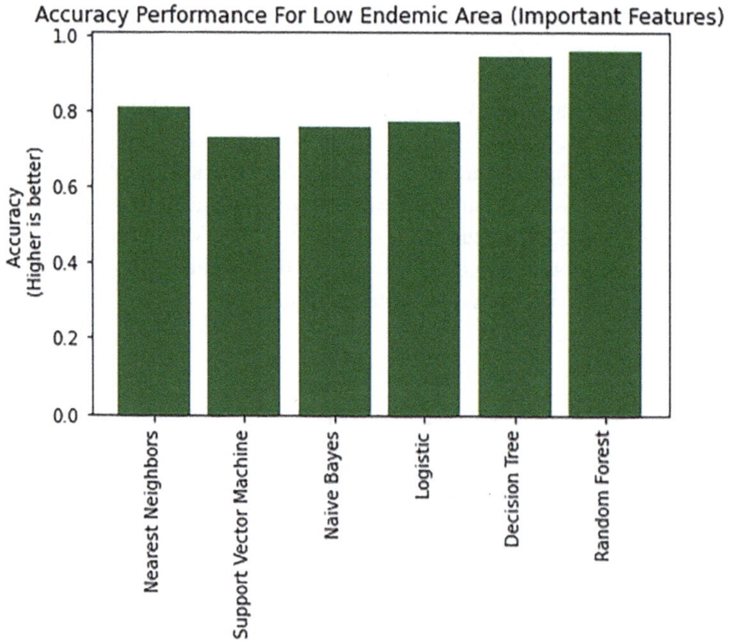

Fig. 5 Accuracy performance for important features in low-endemic dataset

Using the complete set of features, the performance accuracy of the six machine learning classifiers was K-Nearest Neighbour 70%, Support Vector Machine 69%, Naïve Bayes 33%, Logistic Regression 70%, Decision Tree 75% and Random Forest 82%, with an average performance of 70% accuracy. Overall, there is an improvement in the performance accuracy of the model with the datasets that have important features. For example, when a dataset with only important features was used on the machine learning classifiers, the performance accuracy was K-Nearest Neighbour 73%, Support Vector Machine 71%, Naïve Bayes 63%, Logistic Regression 70%, Decision Tree 79%, and Random Forest 86%. The study by [33] in 2022 successfully built a predictive model with important features using the malaria diagnosis dataset.

Healthcare Workers' Perspectives

The study revealed that the main symptoms medical doctors commonly used for malaria clinical diagnosis are fever, shivering, and headache, as shown in Table 3. Furthermore, the doctors use nausea and vomiting, dizziness, loss of appetite, diarrhea, joint pain, limpness, abdominal pain, and heartburn as the supporting symptoms for the clinical diagnosis of malaria. Moreover, according to medical doctors, malaria can also be identified using severe symptoms such as loss of consciousness, anemia, jaundice, enlarged spleen, seizures, and shortness of breath. The study finds that more experienced doctors understand the disease more. Also, severe malaria is clinically presumed when intensity increases among the observed symptoms. The residence area was significant among patients, and knowing the patients' travel history from low- to high-endemic areas is essential in diagnosing malaria.

The medical doctors also identified excessive vomiting, Coca-Cola urine (urine with brown color), confusion, and loss of consciousness as some of the severe. All the doctors agree that age is significant, especially for adults. The doctors also agreed that knowing if there was a family member of the patient who was diagnosed with malaria was important since it could have been for the disease to have been transferred to them by the insect. All the doctors approved seasonal malaria as one of the important features in diagnosing malaria in low- and high-endemic areas. Also, the patient's occupation is important since some work environments are more prone to mosquitos and hence a danger to malaria.

Table 3 Healthcare workers' perspectives

	Doctor 1	Doctor 2	Doctor 3	Doctor 4	Doctor 5	Doctor 6
Working experience	3 years	5 years	7 years	4 years	6 years	5 years
Regions practiced	Kilimanjaro Dar es salaam	Tanga Kilimanjaro Shinyanga	Mwanza Tanga Morogoro	Morogoro Dodoma Shinyanga	Arusha Morogoro Mbeya	Kilimanjaro Dar es salaam Tanga
Main symptoms	Headache Backache Vomiting	Headache Vomiting High fever Body Pain Diarrhea	Fever Headache Vomiting Nausea	Fever Body pain Headache Vomiting	Fever Headache Sweating Body pain	Headache Backache Vomiting
Severe symptoms	Dizziness Anemia Confusion	Excessive vomiting Yellow fever Paleness	High fever Vomiting conversion loss of consciousness Anemia Coca-Cola urine	Confusion fainting Kidney failure	Confusion Loss of conscious	Dizziness Anemia Confusion
Residence area	Yes	Yes (warm areas support parasite growth)	Yes (awareness differs, the use of control initiative differs)	Yes	Yes	Yes
Age	Yes (children under 5)	No	Yes (kids <5)	Yes (<35)	Yes	Yes (children under 5)
Sex	No	Yes (pregnancy)	Yes (pregnancy)	No	No	No
History of travelling	Yes (low prevalence area)	Yes	Yes (from high to low)	Yes	Yes	Yes (low prevalence area)

Using control initiatives	Yes	No	Yes	Yes	Yes	Yes
Family member being sick	Yes	Yes	Yes	Yes	Yes	Yes
Distance from the health facility	Yes	No	Yes	Yes	Yes	Yes
Yearly season	Yes (rainy season)	No	Yes	Yes (planting season)	Yes	Yes (rainy season)
Occupation	Yes (work during the night, farmers)	Yes (e.g., drivers)	Yes	Yes	Yes	Yes (work during the night, farmers)
Malaria prediction Model feasibility	Yes, with concerns	Yes, with concerns	Yes, with concerns	Yes, with concerns	Yes, with concerns	Yes, with concerns
Reasons for Concerns	Substitute the lab confirmation of the disease	Substitute the lab confirmation of the disease	Substitute the lab confirmation of the disease	Substitute the lab confirmation of the disease	Substitute the lab confirmation of the disease	Substitute the lab confirmation of the disease

4 Discussion

For proper disease management, accurate, affordable, and timely diagnosis is vital. In most developing countries, adequate malaria diagnosis has been challenging due to the lack of testing equipment, insufficient personnel to run diagnostic tests, and patients' self-medicating [34–36]. Machine learning can relieve this burden by providing a high-accuracy disease prediction tool that does not require expensive equipment or trained personnel. This can ensure that patients seek treatment at facilities without the recommended equipment or personnel for parasitology tests. Patients visiting pharmacies for medication without testing can be better assessed for the probability of disease before treatment. The accuracy of such prediction models relies on the proper selection of important features for use in training the prediction model. In this study, the aim was to find the most important features in the diagnosis of malaria, and we discovered that not only some of the symptoms have significance in the diagnosis of malaria but also non-symptoms such as the residential area of the patient, sex, and age have significance in the diagnosis of malaria. The difference in the level of important features for different regions signifies that each region is unique even though they are in the same country and should not be treated the same. The difference can be due to geographical location, which can enhance the rate of disease transmission. Apart from the difference in the level of importance, the experiments showed significant features in one region but no significance in the other. For example, coughing and joint pain are significant for malaria diagnosis in Morogoro. Still, they have zero significance in Kilimanjaro, while dizziness and confusion are important in diagnosing malaria in Kilimanjaro and not in Morogoro.

It was also observed that some months of the year when patients visit the health facility with malaria-related symptoms are significant in malaria diagnosis. The significant months are during the rain session or just after the rain session. This aligns with the guideline given by the WHO on malaria transmission behavior. The transmission also depends on climatic conditions that may affect the number and survival of mosquitoes, such as rainfall patterns, temperature, and humidity. In many places, transmission is seasonal, with the peak during and just after the rainy season, as observed in the study done by [29–31, 37, 38]. Malaria epidemics can occur when climate and other conditions suddenly favor transmission in areas where people have little or no immunity to malaria [38]. Both WHO [5, 6] and Tanzania Mainland's malaria treatment guideline [4] propose that for diagnosis of malaria, parasitological confirmation of suspected malaria cases should be given tohfru0-\ patients of all ages with Fever, Headache, Joint pains, Malaise, Vomiting, Diarrhea, Body ache, Body weakness, Poor appetite, Pallor, and Enlarged spleen as a diagnostic criterion. The criteria match the features identified in this study, proving that the model that will be developed will support the malaria treatment guideline. The trained models attained the highest performance accuracy with the selected important features. This means that good feature selection influences a more accurate malaria prediction model.

5 Conclusions

The main objective of this study was to compute the significant features of malaria diagnosis. The results show it is possible to create a more accurate model for malaria prediction by applying feature selection methods to the malaria diagnosis dataset. The ranking of features using a random forest algorithm identified that fever is universally the most noteworthy feature for predicting malaria in all the datasets, followed by general body malaise, vomiting, and headache features; however, these features are ranked differently across the regional datasets. The improvements in performance accuracy over using the original dataset vary greatly depending on which machine learning algorithm is used; therefore, reviewing a wide range of feature selection techniques with machine learning algorithms to get the best possible model is necessary. Also, the important features obtained in this study align with the most significant symptoms and non-symptoms identified by medical doctors. This shows the great feasibility of using this model for malaria diagnosis. This study is limited to selecting the important features for malaria prediction. In future research, looking at different machine learning algorithms for building a malaria predictive model would be interesting. For example, Clinician's pharmacist can use the model built in the further study and different individuals to detect malaria in new patients, provided patient data for the features used are available.

References

1. WHO. (2020). World malaria report 2020. *World Health* Organization. Retrieved May 29, 2021, from https://www.who.int/publications/i/item/9789240015791
2. WHO. (2019). World malaria report 2019. *World Health* Organization. Retrieved May 29, 2021, from https://www.who.int/publications/i/item/9789241565721
3. Chacky, F., et al. (2018, December). Nationwide school malaria parasitaemia survey in public primary schools, the United Republic of Tanzania. *Malaria Journal, 17*(1), 1–16. https://doi.org/10.1186/S12936-018-2601-1
4. Michael, D., & Mkunde, S. P. (2017, April). The malaria testing and treatment landscape in mainland Tanzania, 2016. *Malaria Journal, 16*(1), 1–15. https://doi.org/10.1186/S12936-017-1819-7
5. WHO. (2019). *WHO | Guidelines for the treatment of malaria* (3rd ed.). World Health Organization.
6. WHO. (2021). Introduction - WHO Guidelines for malaria - NCBI Bookshelf. NCBI. Retrieved June 25, 2021, from https://www.ncbi.nlm.nih.gov/books/NBK568497/
7. Budimu, A., Emidi, B., Mkumbaye, S., & Kajeguka, D. C. (2020). Adherence, awareness, access, and use of standard diagnosis and treatment guideline for malaria case management among healthcare workers in Meatu, Tanzania. *Journal of Tropical Medicine, 2020*, 1918583. https://doi.org/10.1155/2020/1918583
8. Hertz, J. T., et al. (2019). Self-medication with non-prescribed pharmaceutical agents in an area of low malaria transmission in northern Tanzania: A community-based survey. *Transactions of the Royal Society of Tropical Medicine and Hygiene, 113*(4), 183–188. https://doi.org/10.1093/trstmh/try138

9. Chipwaza, B., Mugasa, J. P., Mayumana, I., Amuri, M., Makungu, C., & Gwakisa, P. S. (2014). Self-medication with anti-malarials is a common practice in rural communities of Kilosa district in Tanzania despite the reported decline of malaria. *Malaria Journal, 13*(1), 252. https://doi.org/10.1186/1475-2875-13-252
10. Graz, B., Willcox, M., Szeless, T., & Rougemont, A. (2011). Test and treat or presumptive treatment for malaria in high transmission situations? A reflection on the latest WHO guidelines. *Malaria Journal, 10*(1), 1–8. https://doi.org/10.1186/1475-2875-10-136
11. Mwai, L., et al. (2009). Chloroquine resistance before and after its withdrawal in Kenya. *Malaria Journal, 8*(1), 106. https://doi.org/10.1186/1475-2875-8-106
12. Menard, D., & Dondorp, A. (2017). Antimalarial drug resistance: a threat to malaria elimination. *Cold Spring Harbor Perspectives in Medicine, 7*(7), 1–24. https://doi.org/10.1101/cshperspect.a025619
13. Muroa, F., Reyburn, R., & Reyburn, H. (2015, December). Acute respiratory infection and bacteraemia as causes of non-malarial febrile illness in African children: a narrative review. *Pneumonia, 6*(1), 6–17. https://doi.org/10.15172/pneu.2015.6/488
14. Goodyer, L. (2015). Dengue fever and chikungunya: Identification in travellers. *Clinical Pharmacist, 7*(4). https://doi.org/10.1211/cp.2015.20068429
15. Lozano, R., et al. (2020). Measuring universal health coverage based on an index of effective coverage of health services in 204 countries and territories, 1990–2019: A systematic analysis for the Global Burden of Disease Study 2019. *The Lancet, 396*(10258), 1250–1284. https://doi.org/10.1016/S0140-6736(20)30750-9
16. Moreno-Ibarra, M.-A., Villuendas-Rey, Y., Lytras, M. D., Yáñez-Márquez, C., & Salgado-Ramírez, J.-C. (2021). Classification of diseases using machine learning algorithms: A comparative study. *Mathematics, 9*(15), 1817. https://doi.org/10.3390/MATH9151817
17. Fernandez-Lozano, C., et al. (May 2021). Random forest-based prediction of stroke outcome. *Scientific Reports, 11*(1), 1–12. https://doi.org/10.1038/s41598-021-89434-7
18. Jain, D., & Singh, V. (2018). Feature selection and classification systems for chronic disease prediction: A review. *Egyptian Informatics Journal, 19*(3), 179–189. https://doi.org/10.1016/j.eij.2018.03.002. Elsevier B.V.
19. Spencer, R., Thabtah, F., Abdelhamid, N., & Thompson, M. (2020). Exploring feature selection and classification methods for predicting heart disease. *Digit Health, 6*, 2055207620914777. https://doi.org/10.1177/2055207620914777
20. Lee, Y. W., Choi, J. W., & Shin, E. H. (2021, February). Machine learning model for predicting malaria using clinical information. *Computers in Biology and Medicine, 129*, 104151.
21. Han, J., Kamber, M., & Pei, J. (2012). Data preprocessing. In *Data Mining* (pp. 83–124). https://doi.org/10.1016/B978-0-12-381479-1.00003-4
22. Velliangiri, S., Alagumuthukrishnan, S., & Thankumar Joseph, S. I. (2019). A review of dimensionality reduction techniques for efficient computation. *Procedia Computer Science, 165*, 104–111. https://doi.org/10.1016/J.PROCS.2020.01.079
23. Jović, A., Brkić, K., & Bogunović, N. (2015). A review of feature selection methods with applications.
24. Nguyen, C., Wang, Y., & Nguyen, H. N. (2013). Random forest classifier combined with feature selection for breast cancer diagnosis and prognostic. *Journal of Biomedical Science and Engineering, 2013*(05), 551–560. https://doi.org/10.4236/JBISE.2013.65070
25. Genuer, R., Poggi, J. M., & Tuleau-Malot, C. (2010). Variable selection using random forests. *Pattern Recognition Letters, 31*(14), 2225–2236. https://doi.org/10.1016/J.PATREC.2010.03.014
26. Speiser, J. L., Miller, M. E., Tooze, J., & Ip, E. (2019). A comparison of random forest variable selection methods for classification prediction modeling. *Expert Systems with Applications, 134*, 93–101. https://doi.org/10.1016/J.ESWA.2019.05.028
27. Sarker, I. H. (2021). Machine learning: Algorithms, real-world applications and research directions. *SN Computer Science, 2*(3), 160. https://doi.org/10.1007/S42979-021-00592-X

28. Ngasala, B., et al. (2008, October). Impact of training in clinical and microscopy diagnosis of childhood malaria on antimalarial drug prescription and health outcome at primary health care level in Tanzania: A randomized controlled trial. *Malaria Journal, 7*, 199. https://doi.org/10.1186/1475-2875-7-199
29. Chandramohan, D., Carneiro, I., Kavishwar, A., Brugha, R., Desai, V., & Greenwood, B. (2001, July). A clinical algorithm for the diagnosis of malaria: results of an evaluation in an area of low endemicity. *Tropical Medicine and International Health, 6*(7), 505–510. https://doi.org/10.1046/j.1365-3156.2001.00739.x
30. Nkumama, I. N., O'Meara, W. P., & Osier, F. H. A. (2017. Elsevier Ltd). Changes in malaria epidemiology in Africa and new challenges for elimination. *Trends in Parasitology, 33*(2), 128–140. https://doi.org/10.1016/j.pt.2016.11.006
31. Caminade, C., et al. (2014). Impact of climate change on global malaria distribution. *Proceedings of the National Academy of Sciences of the United States of America, 111*(9), 3286–3291. https://doi.org/10.1073/pnas.1302089111
32. Mariki, M., Mkoba, E., & Mduma, N. (2022). Combining clinical symptoms and patient features for malaria diagnosis: Machine learning approach. *Applied Artificial Intelligence, 36*(1), 1–25. https://doi.org/10.1080/08839514.2022.2031826
33. Attinsounon, C. A., et al. (2019). Self-medication practice against malaria and associated factors in the city of Parakou in Northern Benin: Results of a population survey in 2017. *Advances in Infectious Disease, 9*(3), 263–275. https://doi.org/10.4236/AID.2019.93020
34. Belachew Gutema, G., et al. (2011). Self-medication practices among health sciences students: The case of Mekelle University. *Journal of Applied Pharmaceutical Science, 2011*(10), 183–189.
35. Kajeguka, D. C., & Moses, E. (2017). Self-medication practices and predictors for self-medication with antibiotics and antimalarials among community in Mbeya City, Tanzania. *Tanzania Journal of Health Research, 19*(4). https://doi.org/10.4314/THRB.V19I4
36. Chukwuocha, U. M. (2016). Malaria treatment in children based on presumptive diagnosis: A make or mar? *Pediatric Infectious Diseases: Open Access, 01*(02), 6. https://doi.org/10.21767/2573-0282.100006
37. Chandramohan, D., Jaffar, S., & Greenwood, B. (2002). Use of clinical algorithms for diagnosing malaria. *Tropical Medicine and International Health, 7*(1), 45–52. https://doi.org/10.1046/j.1365-3156.2002.00827.x
38. WHO. (2020). Malaria, WHO library. Retrieved October 22, 2020, from https://www.who.int/news-room/fact-sheets/detail/malaria

2-Stage Hybrid Ensemble-Based Heterogeneous Committee Machine for Improving Soil Fertility Status Prediction Performance

Augustine J. Malamsha, Mussa Ally Dida, and Sabine Moebs

1 Introduction

Food security is a major concern not only because of the projected world's population growth that demands for increase in food productivity [1], but also due to variabilities in agricultural factors such as the soil, challenged by deteriorations of its fertility. Essentially, soil fertility validated by its chemical properties has a direct interlink with crop productivity and is often evaluated in agricultural fields across the globe. Despite, the necessity of soil in crop production, yet agricultural soils' susceptibility to low soil fertility characteristics either due to natural or human causes remains a challenge in most of Sub-Saharan Africa (SSA) and South Asia.

Ceteris paribus, inadequate accurate information about the soil is one of the key factors that limit optimal decisions for improving agricultural productivity and food security in general [2]. With the vast amount of data that can be harvested along with agricultural activities, valuable information in the form of trends and patterns can be harnessed for improving crop productivity decision-making. Traditional human or expert methods for gaining such an understanding of vast data are limited, expensive, and may overlook important details and rarely may be subjective and biased. Moreover, classical statistical analysis is often ineffective and inefficient with an increase in the size and complexity of this data.

With the advent of machine learning (ML) techniques, automated data mining tools for analysis, extraction, and delivery of actionable information have been useful in various fields, including financial markets stock prices prediction [3],

A. J. Malamsha (✉) · M. A. Dida
Nelson Mandela African Institution of Science and Technology, Arusha, Tanzania
e-mail: malamshaa@nm-aist.ac.tz

S. Moebs
Baden-Wuerttemberg Cooperative State University, Stuttgart, Germany

J. Marx Gómez et al. (eds.), *Artificial Intelligence Tools and Applications in Embedded and Mobile Systems*, Progress in IS,
https://doi.org/10.1007/978-3-031-56576-2_7

modelling of cholera outbreak, prediction of student drop out rates, and handover management in telecommunication networks [4–6]. In agriculture, these techniques have similarly been potential for solving various knowledge discovery problems ranging from soil analysis and fertility predictions [7], crop disease prediction, to crop monitoring and predicting crop yields. However, varying predictive performances have been demonstrated by ML models for predicting soil fertility status [8]. In addition, a range of varying fertility status levels have been synthesized and used for modelling classifiers targets for predicting soil fertility. For instance, [7] used only two classes "Ideal" and "Not Ideal" as fertility classes "low" and "high," [9] used three classes "low," "medium" and "high," others used up to five fertility classes. While this might have varying implications in application thereof it leaves room for further performance improvement experimentations aiming at the development of reliable high performance throughput ML algorithms models applications.

This paper presents a novel machine learning hybrid model for the classification of soil fertility status using a comprehensive set of soil chemical properties. Specifically, we identify agricultural soil properties to be attributed as features for soil analysis and fertility prediction by using machine learning algorithms, then we use machine learning methods to categorize available distinct soil physiognomies based crop yields as an index to fertility, Finally, we develop a 2-Stage heterogeneous hybrid ensemble committee by using the distinct soil physiognomies to train a diverse of pre-evaluated base machine learning algorithms so as to form an ensemble of heterogeneous experts with an intuition that we could tradeoff the weaknesses of one expert with the strengths of the other experts when that happens, so as to achieve improved prediction performance, by combining the judgments of each expert using a weighted soft voting ensemble scheme of class probability predictions.

The paper is organized as follows, Sect. 2 is related works, Sect. 3 is methodology, and Sect. 4 is results and discussion, and Sect. 5 concludes this paper, and provides suggestions for future work.

2 Related Works

The genesis of ML methods in pedology traces back to 1980s, when it was first applied in pedometrics whereby ML data-driven methods could be applied in the modelling and prediction of soil fertility. Presented here is a briefing on a few previous of these works from 2010 to date (2022).

In [9], a versatile method for rapid and accurate determination of soil fertility for crop production was developed and presented, by using sugarcane as an example, whereby the soil fertility index was established and modelled independently using boosted decision trees where they achieved AUC scores of 0.76, 0.67, and 0.65 for the respective fertility classes prediction. ML RF regression techniques were used to develop a model that was used as part of the work to predict soil nutrient fertilities and use the information to understand the edaphic drivers of yield heterogeneity across Africa, to guide in nutrient-specific interventions [10]. In [11], a hybrid classification model was developed by using a Decision Tree Classifier to isolate dependent features and used Naïve Bayes classification on the independent features [12]. Examined soil micro- and macronutrients using machine learning to grade the examined nutrients and used the graded soil with random forest, support vector machine, and Gaussian Naïve Bayes algorithms to predict the soil classes for suitable crop plantation [13]. implemented machine learning model experiment on the Support Vector Machine, Decision Tree, nearest neighbor, and Naïve Bayes to predict soil fertility as low, high, or medium. It could be observed that some of these studies have single ML algorithm model implementations, while some have used homogenous members in the ensemble committees, and others have featured for binary classification only, and other multi-classification [7]. Utilized data from the Vellore soil testing laboratory to perform experiments of trading various bagging, boosting, and stacking ensemble classifiers, where they pre-processed the data, extracted relevant features as a means to achieve better performance, and attained an accuracy of 98.15% by boosting the decision tree like C5.0 algorithm.

3 Methodology

The study used design science research methodology [14], to identify and pave the objective of the study presented in this paper, ML techniques rigor were the perfect match in solving for both the soil need for adequate and accurate information about soil fertility, and their accuracies is among the key concerns to practitioners.

3.1 Data and Preprocessing

The dataset used to develop the models and associated analysis were primarily obtained from the Tanzania Agriculture Research Institute (TARI), under the African Soil Information Services (AFSIS). Maize yield data was gathered from and Tanzania Ministry of Agriculture Statistical Division. Table 1 shows a description of the dataset for agricultural soil properties and maize yields reference thereof.

3.2 Methods and Experimental Setup

A hybrid of both Unsupervised and Supervised machine learning algorithms were used to implement the required 2-Stage heterogeneous hybrid ensemble committee (2S-HHEC) machine for improving soil fertility status predictive performance experiments, Fig. 1 shows the experimental setup.

Table 1 Description of the dataset for agricultural soil properties and maize yields reference

Attribute	Description	SI unit
OC	Organic carbon	Percentage (%)
pH	Potential hydrogen	
EC	Electrical conductivity	deciSiemens per meter (dS/m)
TN	Total nitrogen	Percentage (%)
P	Phosphorus	Milligrams per kilogram (mg kg^{-1})
Ca	Calcium	Centimoles per kilogram (cmol(+) kg^{-1})
K	Potassium	Centimoles per kilogram (cmol(+) kg^{-1})
Mg	Magnesium	Centimoles per kilogram (cmol(+) kg^{-1})
Na	Sodium	Centimoles per kilogram (cmol(+) kg^{-1})
S	Sulfur	Milligrams per kilogram (mg kg^{-1})
Mn	Manganese	Milligrams per kilogram (mg kg^{-1})
Al	Aluminum	Milligrams per kilogram (mg kg^{-1})
Zn	Zinc	Milligrams per kilogram (mg kg^{-1})
Fe	Iron	Milligrams per kilogram (mg kg^{-1})
B	Boron	Milligrams per kilogram (mg kg^{-1})
M_Yld	Maize yields	Tons per hectare numeric (t/ha)

Data preprocessing. The collected data was pre-processed by separating the top (0–20 cm) instances from those of the subsoil (20–50 cm), removing duplicates, outliers, and extreme values by using the Interquartile Range statistical measure methods for noise filtering and reduction. An aggregation of the soil properties and maize yield data was performed to prepare the resultant data necessary for modelling the machine learning problem.

Fertility Index Derivation and features selection. In the first stage of the 2-Stage hybrid implementation, an additional feature, the fertility class label necessary for the second classification stage of the 2-Stage hybrid was created. This was accomplished through the application of unsupervised machine learning K means algorithm [15] and the automatic knee detection method (K-Elbow) in order to model for the optimal number of fertility classes as characterized in the data, analysis of variance (ANOVA) test was performed on the formed groups to test for groups similarities otherwise differences, with null hypothesis that "the groups are different," this was accomplished through the tukey test. It is necessary to avoid model performance impairment due to complexities that may be caused in many cases by extensive multicollinearity in data [16], thus removal of one the highly correlated features is necessary to remove unnecessarily multicollinearities in data. We, therefore, performed feature selection by identifying and removing features with correlation above 70%, as a standard correlation threshold in most studies.

Model Selection and Weighted Voting Ensemble Optimization. A total of seven algorithms classifiers: support vector machine (SVM) through the SVC, DecisionTreeClassifier (DT), GaussianNB (NB), KNeighborsClassifier (KNN), AdaBoostClassifier (AdaBoost) [15], GradientBoostingClassifier (GB), and RandomForestClassifier (RF) were used in the study experiment with Stratified KFold

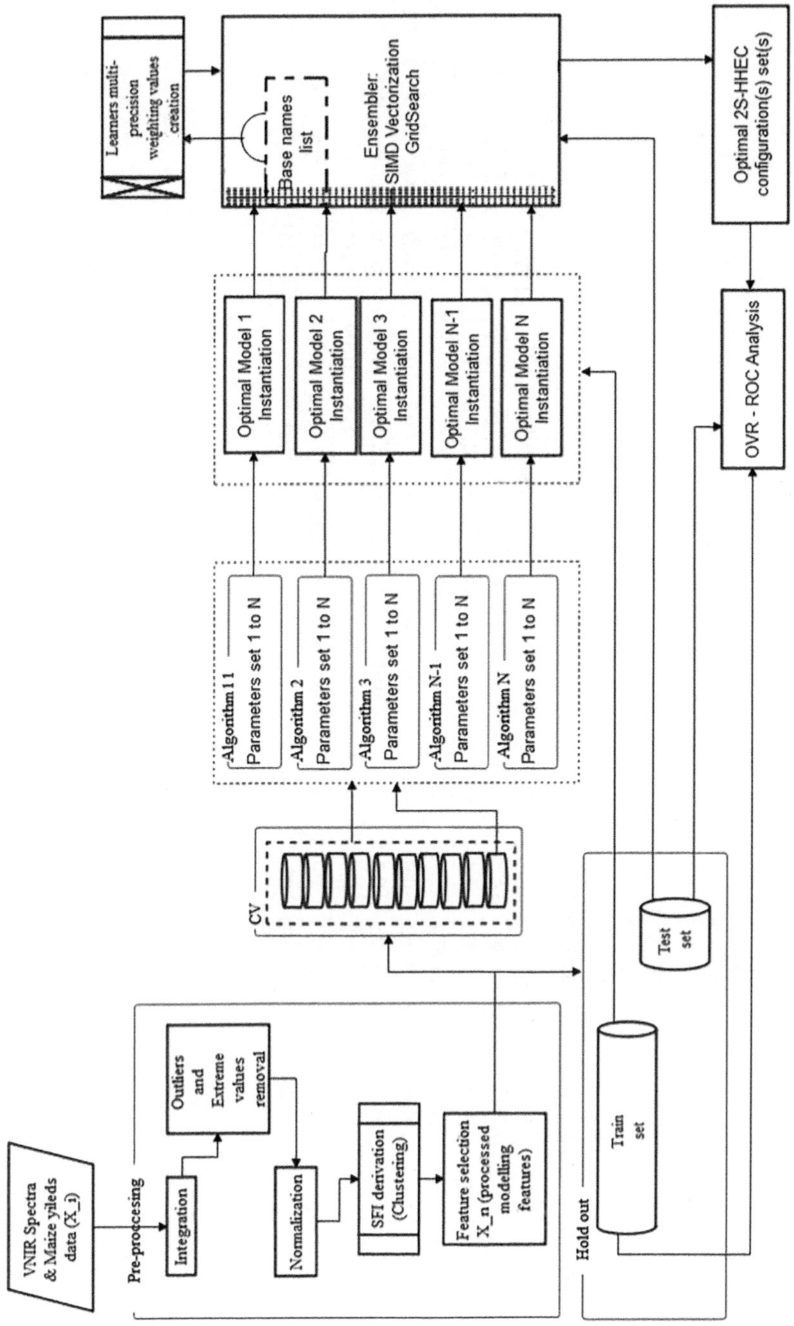

Fig. 1 2S-HHEC experimental setup

Table 2 Statistical description of the used agricultural soils and yield data

	OC	pH	EC	TN	P	Ca	K	Mg	Na	S	Mn	Al	Zn	Fe	B	M_Yld
Count	6260	6260	6260	6260	6260	6260	6260	6260	6260	6260	6260	6260	6260	6260	6260	6260
Mean	1.875	5.912	1.412	0.080	5.003	1.487	0.147	0.180	4.184	56.738	8.391	10.131	0.814	45.689	0.013	1.658
Std	0.804	0.516	2.493	0.043	2.854	1.732	2.450	0.484	22.17	27.196	5.325	29.678	2.314	32.956	0.019	0.679
Min	0.256	4.257	0.001	0.008	0.062	0.005	0.002	0.001	0	0.365	0.006	0.01	0.009	0.401	0.002	0.63
25%	1.211	5.547	1.071	0.044	3.135	0.388	0.02	0.050	0.081	37.616	5.427	0.439	0.476	35.382	0.008	1.22
50%	1.667	5.904	1.368	0.064	4.571	0.960	0.031	0.105	0.207	53.687	7.335	0.566	0.584	41.988	0.011	1.62
75%	2.543	6.274	1.716	0.116	6.27	1.985	0.047	0.200	0.610	73.754	10.312	0.821	0.744	48.895	0.014	2.11
Max	4.211	7.457	139.46	0.206	37.721	27.021	82.313	17.842	803.93	310.55	103.99	554.2	51.441	1068.5	1.003	4.39

Cross Validation (N = 10) on the dataset of size 6260 (see Table 2) for robustly selecting best models across a wide range of their associated tuning parameters through Gridsearch method, The formula in Eq. 1 was implanted on top of a custom automated module for automatic multi-precision weighting values creation to weight the soft voted ensemble committee class probability predictions during optimization to select the best ensemble parameters at even a much bigger precision value.

$$y = \text{argmax} \sum_{i=1}^{N} wi, jXA\,(Cj(x) = j), \tag{1}$$

"where y of all the unknown instances χ in the test sets are evaluated as follows Whereas function argmax return the values of index respective to the large value from array, A = {1, 2, ..., M} denotes the set of exclusive class labels and χA indicates the characteristics function that considered the predictions $j \in A$ of a classifiers Ci on instances and create vectors where the j coordinates take values of one and the remaining takes the value of zero [17]."

Vectorization array programming for SIMD operations was necessary given the nature of the custom an integrated much finer weights values formulation module for automatic multi-precision which may require many computations. Unlike the traditional implementation of Gridsearch that often involves SISD operations, SIMD vectorization provides for fast computations [18]. Hence they are suitable for computations that involve extensive iterations, for that reason it was used to achieve optimality search from the automatically unveiled search grids.

3.3 Performance Evaluation

Measuring the discrimination ability of a model is one of the important aspects in assessing its performance [19]. The metrics we used to evaluate the performance of our models are accuracy, precision, recall, f1-measure, Cohen kappa, and receiver operating characteristics (ROC) analysis, with its associated area under the curve (AUC). The basic performance metric Accuracy, to address the multi-classification problem precision, recall, and f-measure are the most widely used metrics for evaluating the performance of machine learning classifiers. Although past performance may not be indicative of future results, the mentioned metrics formed the common ground for determining how well the developed classifiers might perform in the future. These metrics were computed by using the powerful Scikit-learn (Sklearn) library for ML applications in Python [15], which consists of a wide range of ML classification metrics. While, accuracy of a classifier calculated as the ratio of correct prediction to total observations, is the most spontaneous measure for performance, we used it herein to measure the pseudo baseline performance only for both the base models and the resultant 2S-HHEC, accuracy was not used as the key leading model performance measure.

For detailed investigation of the predictive quality of the resultant 2S-HHEC, the receiver operating characteristics (ROC) analysis was used whereby the measure of its ability to increasingly provide correct predictions could be determined through a 2-dimensional (x, y) plane ROC curves visualizations of its true positive rates (sensitivity) vs false positive rates (1—specificity), in such this helped us determine how correctly each of the "low," "medium," and "high" soil fertility classes may correctly be predicted in future unseen data or new observation. While, ROC curves are useful for a quick visualization of the classifier's quality, area under the curve (AUC) of the ROC (AUC-ROC) was obtained to determine the actual value of the area under ROC curve. Achievements of accuracy close to 100% are desirable, although in imbalanced multi-classification tasks F1 measure under precision and recall are the most suitable measures of classifiers discrimination ability. Cohen kappa was used to measure the inter-rater agreement, Kappa's value closing to one shows great agreement between data collectors, to zero shows no agreement, it was used herein to determine how reliable is it that the model was trained with the right data geometry, i.e., the level of agreement between the variable used to train the model and the data it was supplied with.

4 Results and Discussion

Table 2 shows statistical description of the data used in the experiment showing count, mean, standard deviation (std), min, 25%, 50%, and 75% percentiles, and max for each attribute in the dataset. Results of modelling of Soil fertility

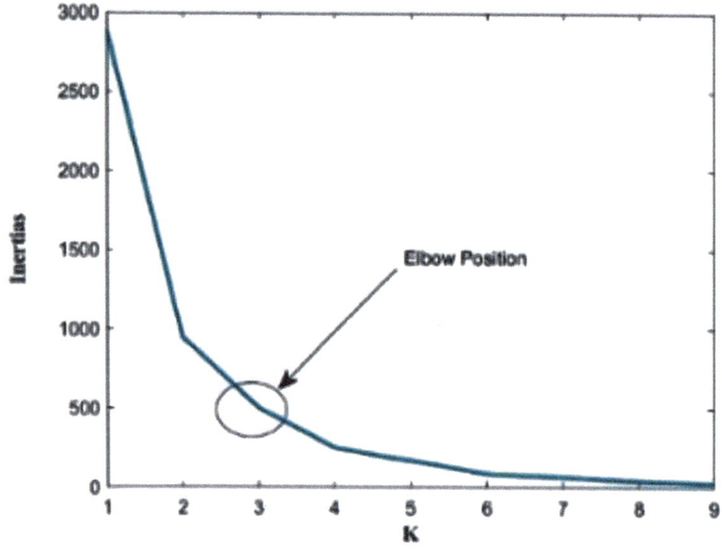

Fig. 2 Knee elbow = 3

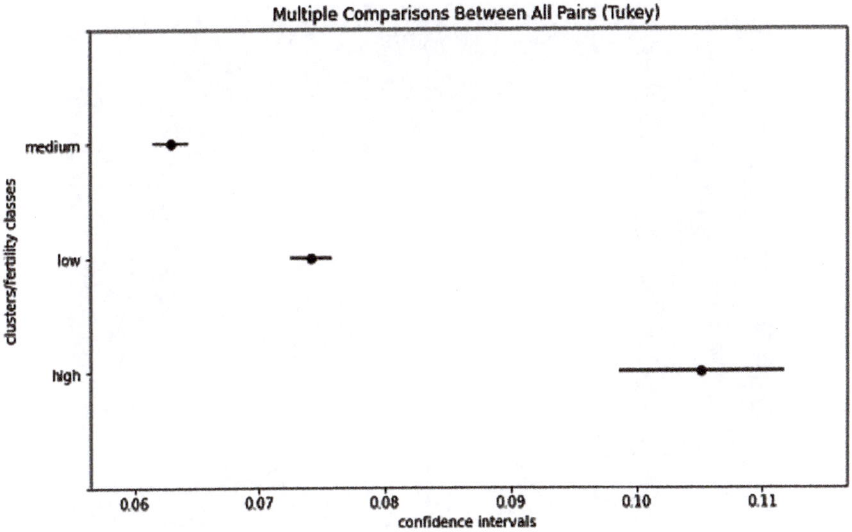

Fig. 3 ANOVA (Tukey Test) results

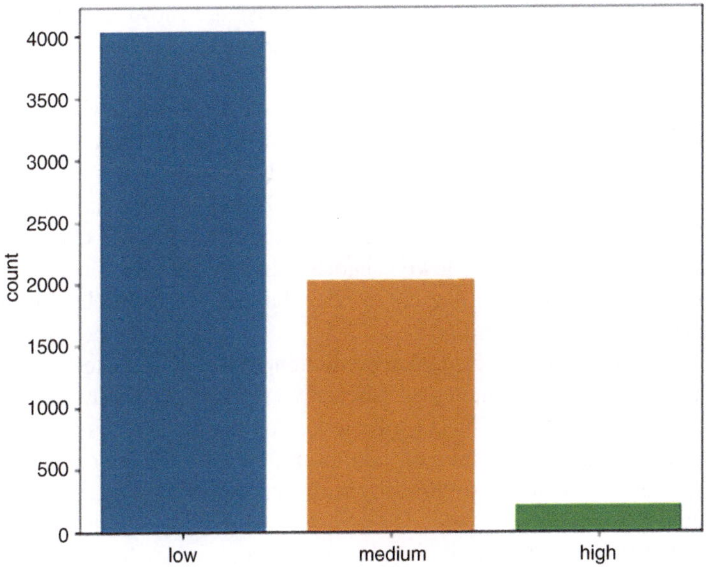

Fig. 4 Soil fertility classes

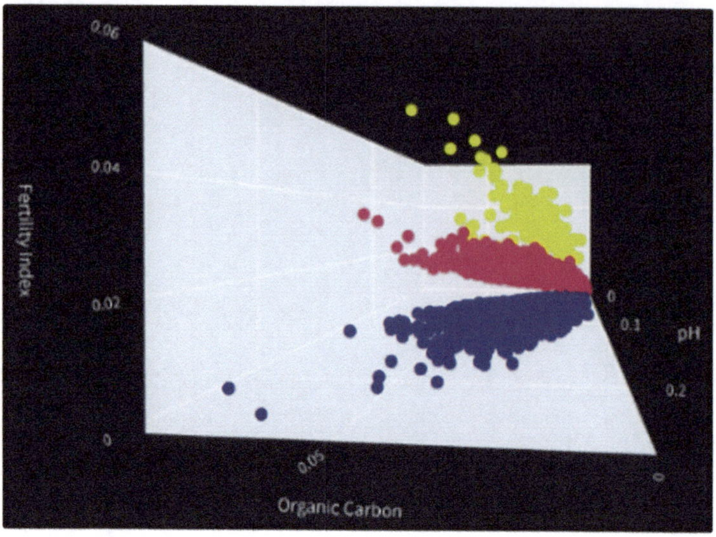

Fig. 5 Organic carbon, pH vs fertility index clusters visualization

characteristic index in the data-preprocessing phase are shown in Figs. 2, 3, 4, 5. Model performances on test data results are displayed in Tables 3 and 4, Figs. 6 and 7.

As shown in Fig. 2, three optimum groups were obtained, which resulted in three distinct soil fertility status indices shown in Fig. 3. ANOVA test results with null hypothesis "the groups are different" tested positive meaning the groups are different as shown in Fig. 3, for the respective pH and Organic carbon, vs fertility index dimension. Figure 4 shows the distribution of the fertility classes. It could be seen that major parts are lowly characterized by soil fertility, these results match with the results of the authors in [20], that 80% of arable land in Africa is characterized by low soil fertility.

The base model demonstrated good prediction performance on test data, gradient boosting classifier scoring the highest accuracy with 93%, respective to 91%, 91%, and 90% for RF, SVM, and KNN (see Table 3). Comparing these results with similar objective model accuracy results of authors in [13] who implemented a Decision Tree classifier for predicting soil fertility with an accuracy of 60%, our results were 30% better.

GB, RF, SV, KN, DT hybrid ensemble committee could achieve an accuracy of 98.93%, a score higher by 4.03% than result of authors in [7]. However, it could be observed that, both the new model and benchmark had a Kappa score close to one, indicating that both models had good inter-rater reliability indicating high agreement, although the benchmark has 1% higher as much agreement than the new model (see Table 4). Visual inspection of the ROC curve of the new model in Fig. 7, it can be seen from the respective ROC curves for high, medium, and low are all above the 0.5 cut point, indicating the model is performing reasonably

Table 3 Base models performance

Model	Accuracy	Precision				Recall				F1_Score			
		Low	Mid	High	Weighted	Low	Mid	High	Weighted	Low	Mid	High	F1_Score
KNN	0.90	0.82	0.97	0.62	0.91	0.93	0.88	1	0.9	0.87	0.92	0.76	0.9
SVM	0.91	0.85	0.93	0.86	0.91	0.86	0.92	0.98	0.91	0.86	0.93	0.91	0.91
DT	0.87	0.79	0.94	0.56	0.88	0.87	0.87	0.9	0.87	0.83	0.9	0.69	0.87
RF	0.91	0.89	0.94	0.66	0.92	0.87	0.93	0.97	0.91	0.88	0.94	0.78	0.92
AdaB	0.52	0.43	0.75	0.1	0.63	0.48	0.53	0.6	0.52	0.45	0.62	0.17	0.55
NB	0.11	0.3	0.81	0.04	0.62	0.13	0.06	0.9	0.11	0.18	0.11	0.07	0.13
GB	0.93	0.89	0.95	0.82	0.93	0.89	0.94	0.97	0.93	0.89	0.95	0.89	0.93

Table 4 New 2S-HHEC based and second benchmark model performance

	New model(s)		Benchmark(s)							
	Weighted voting		Bagging		Boosting		Stacking			
	Gb,RF,SV,KN,DT	Gb,RF,SV,KN	TreeBag'	RF	C5.0	Gbm	LR	KNN	CRT	SVM
Accuracy	98.93	93.98	94.4	96.3	98.15	92.95	73.82	78.43	73.25	92.22
Kappa	93.9	87.2	88.06	92.24	94.9	76.98	48.01	30.58	19.5	79.36
AUC (micro)	0.92	0.92	–	–	–	–	–	–	–	–
AUC (macro)	0.83	0.83	–	–	–	–	–	–	–	–

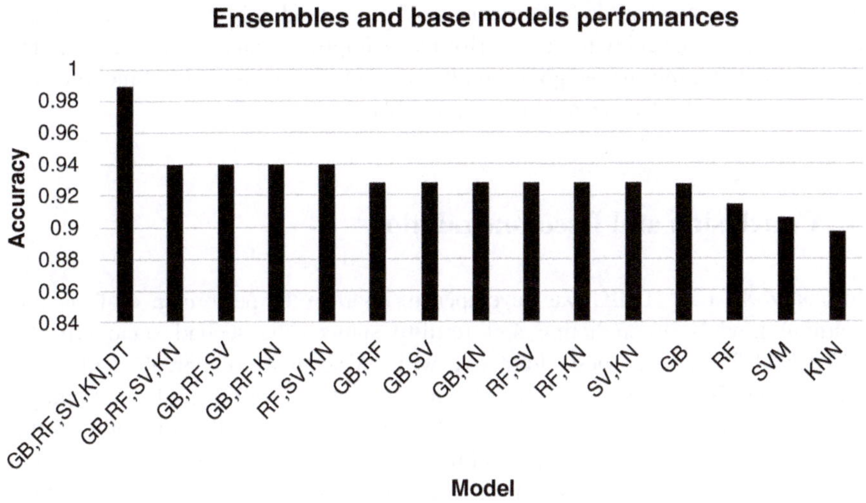

Fig. 6 2S-HHEC's and learners performances

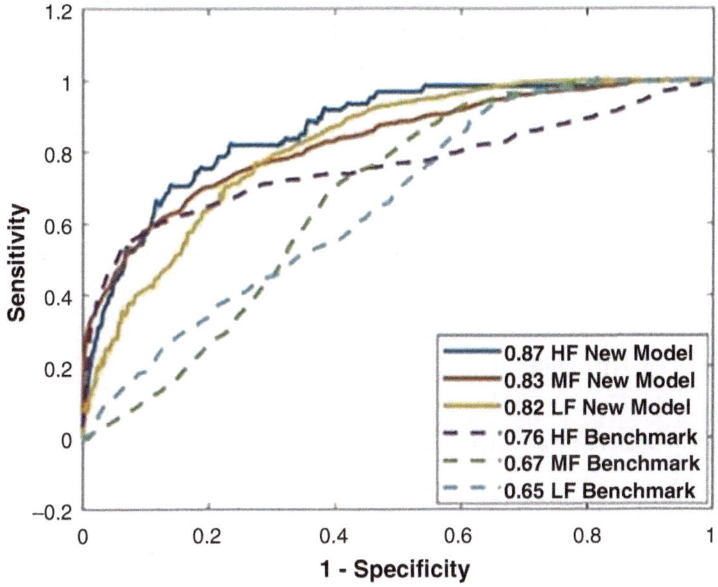

Fig. 7 ROC curves for New vs Benchmark

good in correctly predicting each of those classes with actual AUC values of 0.87, 0.83, and 0.82. These results had an improvement compared to the authors in [9]. From the ROC curves, it could be observed that the new model is above reasonably performing well with an increasing ability to provide correct predictions, with cut point 0.5, it is not bad and extremely far from a randomly guessing model (see Fig. 7). Success in achieving such performance improvements was also as part of an effort of automating the weighting multi-precision values mapping function as an assurance to exhaustive search in the optimality space.

5 Conclusion and Recommendations

In this work, a 2S-HHEC was developed to improve the performance of machine learning models in predicting soil fertility status. The second stage GB, RF, SV, KN, DT were the best alternative with an accuracy of 98.93% and Cohen Kappa 93.98% on test data. The presented 2S-HHEC effort could be deployed for use by soil testing laboratories and farmers in order for them to gain a better understanding of farm fertility conditions through accurate predictions, leading to the application of both site-specific and fine-tuned fertilizer dosages, as well as the appropriate management interventions necessary to ensure sustainability in soil fertility management as a key to sustainable agricultural intensification for ensuring food security for the world growth, which is estimated to reach 8 billion people by 2030, and 10 billion by 2050 [1]. Future work, maybe to explore the applications of the Artificial Intelligence (AI) genetic evolutionary Algorithm's (GA) to generate populations of classification models, with an intuition to evolve among them, the fittest predictor with survivability in major sorts of weather and climatic uncertainties.

Acknowledgments The work has been supported by the African Development Bank (AfDB), United Republic of Tanzania, through project No.: P-Z1-IA0-016 and grant No.: 2100155032816.

Conflicts of Interest

The authors declare no conflicts of interest.

References

1. FAO. (2018). *Future of food and agriculture 2018: Alternative pathways to 2050*. Food & Agriculture Org.
2. Massawe, B. H., Subburayalu, S. K., Kaaya, A. K., Winowiecki, L., & Slater, B. K. (2018). Mapping numerically classified soil taxa in Kilombero Valley, Tanzania using machine learning. *Geoderma, 311*, 143–148.
3. Guo, J.-J., & Luh, P. B. (2004). Improving market clearing price prediction by using a committee machine of neural networks. *IEEE Transactions on Power Apparatus and Systems, 19*(4), 1867–1876.

4. Mduma, N. (2020). *Data driven approach for predicting student dropout in secondary schools.*," PhD Thesis,. NM-AIST.
5. Leo, J., Luhanga, E., & Michael, K. (2019). Machine learning model for imbalanced cholera dataset in Tanzania. *Scientific World Journal, 2019*, 9397578.
6. Mollel, M., et al. (2019). Handover management in dense networks with coverage prediction from sparse networks. In *2019 IEEE Wireless Communications and Networking Conference Workshop (WCNCW)* (pp. 1–6).
7. Jayalakshmi, R., & Savitha Devi, M. (2022). Mining agricultural data to predict soil fertility using ensemble boosting algorithm. *International Journal of Information Communication Technologies and Human Development IJICTHD, 14*(1), 1–10.
8. Janvier, N., Arcade, N., Eric, N., & Jean, N. (2021). Machine learning based soil fertility prediction. *International Journal of Innovative Science, Engineering & Technology, 8*(7), 5.
9. Viscarra Rossel, R. A., Rizzo, R., Demattê, J. A. M., & Behrens, T. (2010). Spatial modeling of a soil fertility index using visible–near-infrared spectra and terrain attributes. *Soil Science Society of America Journal, 74*(4), 1293–1300.
10. Jin, Z., et al. (2019). Smallholder maize area and yield mapping at national scales with Google earth engine. *Remote Sensing of Environment, 228*, 115–128.
11. Manjula, E., & Djodiltachoumy, S. (2017). Data mining technique to analyze soil nutrients based on hybrid classification. *International Journal of Advanced Research in Computer Science, 8*(8).
12. Keerthan Kumar, T. G., Shubha, C., & Sushma, S. A. (2019). Random forest algorithm for soil fertility prediction and grading using machine learning. *International Journal of Innovative Technology and Exploring Engineering, 9*(1), 1301–1304.
13. Chaudhari, R., Chaudhari, S., Shaikh, A., Chiloba, R., & Khadtare, T. D. (2020). Soil fertility prediction using DM. *Bulletin Monumental, 21*(01), 8.
14. Gregor, S., & Hevner, A. R. (2013). Positioning and presenting design science research for maximum impact. *MIS Quarterly, 37*(2), 337–355.
15. Pedregosa, F., et al. (2011). Scikit-learn: Machine learning in python. *Journal of Machine Learning Research, 12*, 2825–2830.
16. Alin, A. (2010). Multicollinearity. *Wiley Interdisciplinary Reviews: Computational Statistics, 2*(3), 370–374.
17. Escorcia-Gutierrez, J., Gamarra, M., Soto-Diaz, R., Pérez, M., Madera, N., & Mansour, R. F. (2022). Intelligent agricultural modelling of soil nutrients and pH classification using ensemble deep learning techniques. *Agriculture, 12*(7), 977.
18. Raskulinec, G. M., & Fiksman, E. (2015). SIMD functions via OpenMP. In *High performance parallelism pearls* (pp. 421–440). Elsevier.
19. Pearce, J., & Ferrier, S. (2000). Evaluating the predictive performance of habitat models developed using LR. *Ecological Modelling, 133*(3), 225–245.
20. Hengl, T., et al. (2015). Mapping soil properties of Africa at 250 m resolution: Random forests significantly improve current predictions. *PLoS One, 10*(6), e0125814.

Convolutional Neural Network Deep Learning Model for Early Detection of Streak Virus and Lethal Necrosis in Maize: A Case of Northern-Highlands, Tanzania

Flavia Stephen Mayo and Neema Mduma

1 Introduction

Globally, plant diseases are regarded as the major issue facing the agricultural industry [1]. Regardless of geographical locations and size of operations of the agricultural experts including farmers and plant pathologists, the common issues faced are: plant infections, plant stress, and pest attacks [2–4]. Farmers must be promptly informed of the presence of such issues in their farms. However, it would be very beneficial for agricultural experts to have exposure to widely available technologies that could teach them how to address each of these threats to agricultural productivity in order to increase crop yield and other products profit [1].

Most plant leaf diseases result in deterioration in the quality and quantity of agricultural products [5]. For the Tanzanian context, food and cash crops are the main categories of agriculture. The food crop is dominated by maize crop, which is primarily used as food, and a significant crop accounting for about 45% of the country's farmland, it moreover provides a mean earning of 100 USD per family per annum in remote areas. The crop is commonly used to prepare stiff porridge (ugali) which is a staple food within the country [6]. With that being the case, yet maize crop production is challenged with numerous diseases, that jeopardize the nation's food security. Nevertheless, several researches are ongoing to ensure that agricultural production is enhanced in meeting the current and future food demands [7].

F. S. Mayo (✉) · N. Mduma
Computational and Communication Science and Engineering (CoCSE), The Nelson Mandela African Institution of Science and Technology (NM-AIST), Arusha, Tanzania
e-mail: mayof@nm-aist.ac.tz

J. Marx Gómez et al. (eds.), *Artificial Intelligence Tools and Applications in Embedded and Mobile Systems*, Progress in IS,
https://doi.org/10.1007/978-3-031-56576-2_8

According to [8, 9], the maize crop is highly affected by Maize Lethal Necrosis (MLN). Moreover, [10–12] added that Maize Streak Virus (MSV) is also a major hindrance to maize crop production. These two diseases pose a severe threat to maize output in East Africa and Sub-Saharan Africa in general [13]. Early detection and control of these diseases are vital in increasing productivity, still less information is available on a combined deep model for detection of the diseases [14, 15].

Currently, the technique employed in disease identification by most agricultural experts is human visual inspection. Unfortunately, it is difficult to determine the condition of crops by visual assessment without wasting time, money, and accuracy. To address this issue, researchers devised a number of solutions, among them being deep learning models that are being used to generate new technologies such as object detection and image processing for quality ratings. Over the last few years, Deep learning (DL) has achieved significant advances in computer vision and object recognition. It became accustomed to successfully identify, categorize, and quantify a wide range of infections, pests, and stress in a variety of crops [7, 16]. There are several deep learning models in place for the detection of plant diseases, but the Convolutional Neural Network (CNN) model is proposed as the best suit for identifying diseases in maize and banana crops [17–19].

Most conducted studies have been developing solutions using image data that have not been sufficiently curated, prepared and shared within the wider machine learning community. Since there is a lack of publicly available datasets to facilitate machine learning activities in Africa, this study proposes to generate a dataset for maize leaf images and develop a deep learning model for early detection of maize diseases.

2 Materials and Methods

2.1 Experimental Setup

The experiment was conducted on a machine running Windows 10 with an Intel(R) Core (TM) i5-4200U CPU @ 1.60 GHz 2.30 GHz having an installed RAM of 8 GB with a system type of 64-bit operating system. We trained the model online using Google Collab, consisting of Python3 as the run-time and GPU as the hardware accelerator. The language used during model training was Python, the implementation was carried out using Keras library with TensorFlow on the backend.

2.2 Dataset

In this work, maize images were collected from small-scale farmers in Arusha, Kilimanjaro, and Manyara regions for 6 months from February to July 2021. We used different mobile phone cameras with different resolutions and images were in a Joint Photographic Group (JPG) format. A total of 18,304 images were collected and distributed to three class labels. Healthy 6072 images, MLN 5373 images and MSV 6859 images. We managed to have a dataset that meets our exact specifications of different images, as shown in Fig. 1.

2.3 Data Cleaning

One of the most important steps in the beginning of every machine learning project is data cleaning. In this stage, we are making sure that the dataset is free of inaccurate or false data. It may be carried out either manually using data wrangling tools or automatically using a computer software. In this step, duplicate images were removed using VisiPics tool, images were labelled according to their respective classes such as HEALTHY_1.jpg to last number, MSV_1.jpg to the last number, MLN_1.jpg to the last number using a Bulk rename utility tool and lastly resized to a uniform pixel of 512*512. Table 1 shows the total number of images before and after duplicates.

a b c

Fig. 1 Sample images from maize datasets where (**a**). Healthy, (**b**). MLN, and (**c**). MSV

Table 1 Total number of images before and after removing duplicates

Classes	Number of images before duplicates	Number of images after duplicates
Healthy	6072	5542
MLN	5373	5347
MSV	6859	6668
Total	18,304	17,557

Table 2 Hyperparameters are used for training the model

Parameters	Value
Epoch	50
Batch size	32
Optimizer	Adam
Metric	Accuracy
Losses	Sparse categorical Crossentropy

2.4 Training and Testing Datasets

A Convolutional Neural Network (CNN) model was developed from scratch using a total amount of 1500 datasets belonging to three classes (HEALTHY, MLN and MSV). The dataset was split into 80% for training, 10% for validation and 10% for testing. Ten percent of validation data was used at the end of every epoch to do validation. Ten percent of test data was used after the 50th epoch to measure the accuracy of the model and also test the performance of the model before deploying. We used a sequential model which we defined six convolutional layers each followed by a max pooling layer. The first convolutional layer had 32 filters, the second layer up to the sixth layer had 64 filters each, followed by a flatten layer then dense layer with 64 neurons. All convolutional layers used a Rectified Linear Unit (ReLu) as an activation function. The output dense layer had three neurons which is the number of our classes and softmax activation. The images were rescaled by (1.0/255) and they were resized to 512 x 512 pixels. Training dataset was expanded using data augmentation by performing random image flipping horizontally and vertically, and random image rotation of 0.2. Furthermore, hyperparameters were used for training the model as shown in Table 2.

3 Results

After training the model, the results for training accuracy, training loss, validation accuracy and validation loss were observed during training, from the first epoch to the 50th epoch as summarized in Table 3 and plotted in Fig. 2.

The model was then tested using the remaining 10% of testing set and the results for accuracy were 94.79% and Loss was 14.12% are shown in Table 4.

Table 3 The results for training the model

Model results	Value
Training accuracy	95.9%
Training loss	12.2%
Validation accuracy	98.44%
Validation loss	6.33%

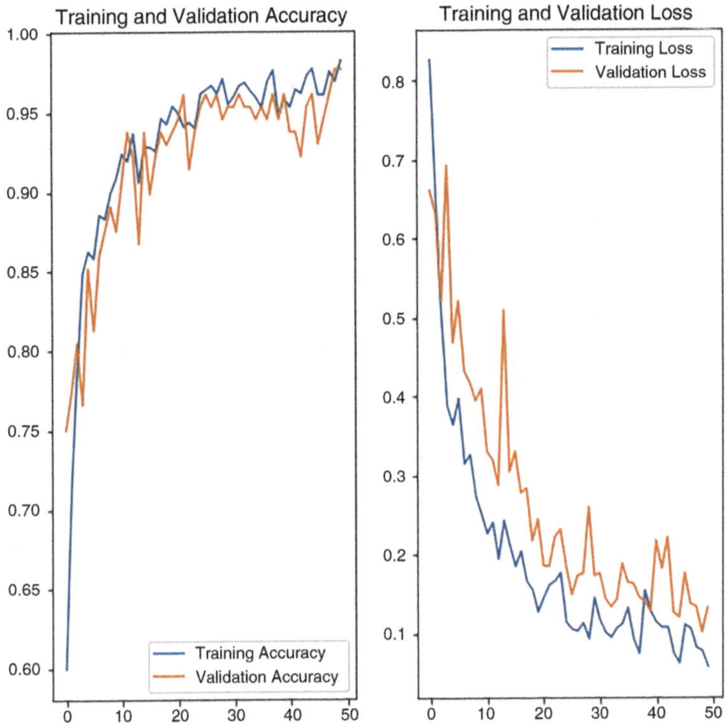

Fig. 2 Training and validation plots for CNN model

Table 4 The results for testing the model

Results	Value
Accuracy	94.79%
Loss	14.12%

4 Discussion

The obtained CNN model accuracy of 95.9% in this study is considered in a group of best model accuracy since a good model is supposed to have an accuracy of above 70% [20]. An accuracy of 95.9% implies that out of 100% samples tested, there is 95.9% correct prediction of maize diseases. Similar results were obtained by Syarief and Setiawan, whereby seven different CNN model architectures were tested with an accuracy of 95% [18]. Hence, this model is reliable and can adequately be used for early detection of maize diseases.

5 Conclusion

In this study, we developed a CNN model for the detection of Maize Streak Virus and Maize Lethal Necrosis. The model will facilitate early detection of maize diseases that can help farmers find better ways of handling them before being severely destroyed. In so doing, the agricultural custodians being the Ministry of Agriculture, companies and other organizations would be able to early detect maize diseases, hence raising the national Gross Domestic Product (GDP) and ensuring food security in the nation. In the future, we plan to develop vision transformer (ViT) model in order to compare with the available CNN model. The best-performed model will then be deployed in a mobile-based application in order to allow easy interactions of the developed model with the farmers, agricultural experts and plant pathologists.

Acknowledgement This research is entirely funded by the International Development Research Centre (IDRC) and Swedish International Development Cooperation Agency (SIDA) through Lacuna Fund in Agriculture and African Centre for Technology Studies (ACTS).

References

1. Ahmed, A. A., & Harshavardhan Reddy, G. (2021). A mobile-based system for detecting plant leaf diseases using deep learning. *Agri Engineering, 3*(3), 478–493.
2. Sinha, A., & Shekhawat, R. S. (2020). Review of image processing approaches for detecting plant diseases ISSN 1751-9659. *IET Image Processing, 14*(8), 1427–1439.
3. Wang, B. (2022). Identification of crop diseases and insect pests based on deep learning. *Scientific Programming, 2022*, 1–10.
4. Zeng, Q., Ma, X., Cheng, B., Zhou, E., & Pang, W. (2020). GANS-based data augmentation for citrus disease severity detection using deep learning. *IEEE Access, 8*, 172882–172891.
5. Patil, M. A. N., & Pawar, M. V. (2017). Detection and classification of plant leaf disease. *Iarjset, 4*(4), 72–75.
6. Frederick, B. et al. (2020). Maize production manual for smallholder farmers in Tanzania, p. 32.
7. Panigrahi, K. P., Das, H., Sahoo, A. K., & Moharana, S. C. (2020). Maize leaf disease detection and classification using machine learning algorithms. In *Progress in computing, analytics and networking* (pp. 659–669).
8. Mahuku, G., et al. (2015). Maize lethal necrosis (MLN), an emerging threat to maize-based food security in sub-Saharan Africa. *Phytopathology, 105*(7), 956–965.
9. Ngaboyisonga, C., Uzayisenga, B., & Gafishi, M. K. (2014, August). Maize Lethal Necrosis (MLN) disease threatens maize production in Rwanda.
10. Charles, K. (2014). Maize streak virus: A review of pathogen occurrence, biology and management options for smallholder farmers. *African Journal of Agricultural Research, 9*(36), 2736–2742.
11. Emeraghi, M., Achigan-Dako, E. G., Nwaoguala, C. N. C., & Oselebe, H. (2021). Maize streak virus research in Africa: An end or a crossroad. *Theoretical and Applied Genetics, 134*(12), 3785–3803.
12. Mbong, G. A., et al. (2021). Phytosanitary situation of Maize streak virusin the main maize production zones of cameroon. *Agricultural Sciences, 12*(04), 339–353.

13. Boddupalli, P., et al. (2020). Maize lethal necrosis (MLN): Efforts toward containing the spread and impact of a devastating transboundary disease in sub-Saharan Africa. *Virus Research, 282*, 197943.
14. Sanya, W., Chande, H., & Haji, H. A. (2021). Maize lethal necrosis disease detection for maize crop real-time prediction yield modeling through colour pixel feature. *International Journal of Computers and Applications, 183*(18), 1–9.
15. Darwish, A., Ezzat, D., & Hassanien, A. E. (2020). An optimized model based on convolutional neural networks and orthogonal learning particle swarm optimization algorithm for plant diseases diagnosis. *Swarm and Evolutionary Computation, 52*, 100616.
16. Singh, A., Subramanian, B. G., Singh, A. K., & Sarkar, S. (2016). Machine learning for high-throughput stress phenotyping in plants. *Trends in Plant Science, 21*(2), 110–124.
17. Sibiya, M., & Sumbwanyambe, M. (2021). Automatic fuzzy logic-based maize common rust disease severity predictions with thresholding and deep learning. *Pathogens, 10*(2), 1–17.
18. Syarief, M., & Setiawan, W. (2020). Convolutional neural network for maize leaf disease image classification. *Telkomnika (Telecommunication Computing Electronics and Control), 18*(3), 1376–1381.
19. Sanga, S. L., Machuve, D., & Jomanga, K. (2020). Mobile-based deep learning models for Banana disease detection. *Engineering Technology and Applied Science Research, 10*(3), 5674–5677.
20. Maxwell, A. E., Warner, T. A., & Guillén, L. A. (2021). Accuracy assessment in convolutional neural network-based deep learning remote sensing studies—Part 1: Literature review. *Remote Sensing, 13*(13), 2591.

Object Detection Model for Poultry Diseases Diagnostics

Shabani Karim, Dina Machuve, Elizabeth Mkoba, and Sanket Pandhare

1 Introduction

1.1 Background

Poultry production has become more popular recently among many households thanks to its considerable income source and superior protein suitable for the current increase in population and urbanization [1]. Tanzania now has over 35 million chickens, of which about 33 million are local village chickens kept primarily in rural regions, and about 2 million are broiler and layer chickens raised by both smallholder farmers and large-scale farms [2]. Due to its ease of adaptation to numerous microhabitats [3], chicken, particularly indigenous chicken, is becoming the most commonly kept livestock species in both rural and peri-urban settings [4]. Other poultry species include pigeons, ducks, turkeys, and guinea fowl. According to the *2002/2003 National Sample Census of Agriculture* [5], 3,017,004 (62%) of Tanzania's 4,901,837 households kept chickens; 99% of these kept native chickens, and the other 1% kept commercial chickens.

However, the pace of poultry population growth is moderate, with an average yearly rise of 2.6% in Tanzania mainland [6]. Besides inadequate nutrition, housing, and management, one of the most significant obstacles facing the chicken industry worldwide is disease [7–9]. Diseases lead to decreased chicken production, which results in financial losses, increased veterinary costs, culling, and actual chicken

S. Karim (✉) · E. Mkoba · S. Pandhare
Nelson Mandela African Institution of Science and Technology, Arusha, Tanzania
e-mail: mdungus@nm-aist.ac.tz

D. Machuve
DevData Analytics, Arusha, Tanzania

J. Marx Gómez et al. (eds.), *Artificial Intelligence Tools and Applications in Embedded and Mobile Systems*, Progress in IS,
https://doi.org/10.1007/978-3-031-56576-2_9

mortality. Generally speaking, the ongoing persistence of diseases has a significant negative influence on food security, health, nutrition, and financial costs at different levels, including producer, local, municipal, and national economies [10]. Among many diseases that are affecting poultry, most of them can be categorized as either infectious or non-infectious and a few are caused by supplement deficiencies. The most infectious common diseases include Newcastle Disease, Fowl Pox, Salmonellosis, Pullorum, and the Non-Infectious are rickets, curled toe paralysis, and encephalomalacia [11]. When it comes to poultry diseases, the flock must be taken into account rather than specific birds. Typically, symptoms in a small number of birds are a sign of a more significant flock-wide issue [11]. Counting the number of oocytes in feces or intestinal scrapings, isolation and identification, and PCR processes, which are used for the diagnosis of multiple diseases, are just a few of the methods used in disease diagnostics in chickens [12]. The main methods of disease transmission are oral via the navel, contaminated feed, and excretions from diseased chicken. The diseased chicken may exhibit respiratory or intestinal symptoms [12]. This work aims to use data-driven approaches to detect three poultry diseases namely, Newcastle disease, Salmonella, and Coccidiosis.

Images are artifacts that represent visual perception. They have been utilized for detection and diagnosis in a variety of disciplines, including health, agriculture, and many other fields [13, 14]. Deep learning, one of the key branches of machine learning, has proven to be quite effective at classifying images. Based on a variety of factors, including color, different levels of characteristics are recorded and assessed. Typically, the way features are presented in the images affects classification quality. The digestive clinical sign of chicken infected with coccidiosis is severe blood/brown diarrhea; salmonella is white diarrhea [12]. As a result, a data-driven method of image classification is far more robust to a wide variety of image features and diseases. This work focuses on extending the image classification work on early detection of poultry diseases to using object detection. The object detection task will show where the damage is on the poultry fecal which will improve the explainability of the given models to both the poultry farmers and the agricultural experts.

1.2 Overview of Object Detection Approach

Detecting instances of semantic objects of a certain class, such as people, buildings, cars, or animals, in digital images and videos are the task of object detection, a branch of computer science connected to computer vision and image processing. Detecting objects involves two steps: Object Localization which is then followed by Object Classification. This is done by predicting the bounding box's coordinates and the class probabilities [15].

Object detection task mainly uses the YOLO framework. An image and the actual values for the bounding boxes are provided as input. The entire input image is split into a square grid, and the object is detected in the grid cell that contains the object's center. B bounding boxes and their accompanying confidence scores will be predicted for each grid cell. This score reflects the box's accuracy and the model's level of confidence that the object is inside. The confidence score is the Intersection over Union (IoU) of predicted values and the ground truth values of the bounding box [16].

YOLOv4 is a SOTA (state-of-the-art) real-time Object Detection model [16]. It is the fourth YOLO installment and was released by Alexey Bochkovsky [17] in April 2020. It attained SOTA performance On the COCO dataset, which contains 80 various object categories. YOLO is a one-stage detector. The One-stage method, which prioritizes inference speeds, is one of the two primary state-of-the-art approaches utilized for the task of object detection. In one-stage detector models, ROI (Region of Interest) is not selected, the classes and the bounding boxes for the complete image are predicted. Thus, this makes them faster than two-stage detectors [16]. Other examples are FCOS, RetinaNet, and SSD. It separates the object-detection task into two phases: regression and classification. Regression aids in locating objects by predicting classes and bounding boxes for the entire image in a single run. The class of an object is established by classification.

YOLOv5 which is a family of compound-scaled object detection models was developed using the COCO dataset and includes basic Test Time Augmentation functionality (TTA), model ensembling, hyperparameter evolution, export to ONNX, CoreML, and TFLite [18]. It comes in various versions, each having its own unique characteristics. The versions are named: (1) Yolov5s—the small version, (2) Yolov5m—the medium version, (3) Yolov5l—the large version, and (4) Yolov5x—the extra-large version. Glenn Jocher's [19] performance analysis of each of these models is shown in Fig. 1 below.

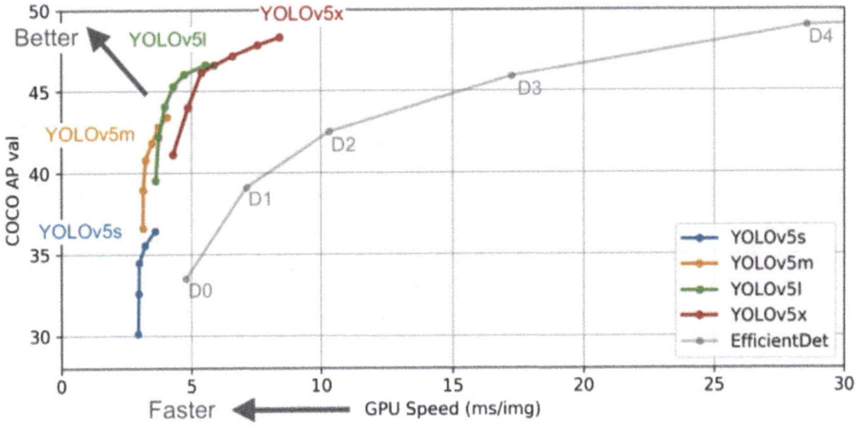

Fig. 1 A comparative plot of performance of the Yolov5 family (Source: https://github.com/ultralytics/yolov5)

2 Methodology

The poultry detection model was trained on YOLOv5. The biggest advantage of using YOLO approach is its superb speed—it is incredibly fast and can process 45 frames per second. YOLOv5 can perform object detection tasks with state-of-the-art results and is quick and simple to use. It is a preferred option for many developers because it is more accurate and simpler to train than its predecessors [19]. YOLO also understands generalized object representation. To start off with YOLOv5, we first cloned the YOLOv5 repository (https://github.com/ultralytics/yolov5) and installed dependencies. By doing this, we will prepare our programming environment for the execution of commands for object detection training and inference. To train Yolov5, we followed the steps indicated below:

- Data collection: During data collection phase there were two main actors that are instrumental to the accurate and timely collection of data, the veterinarian and the field officer. The veterinarian is the expert on poultry diseases, and not only would they help us collect the diseased and healthy classes, but they would also later provide advice and free services to the farmers we visited. The field officer provides up-to-date information on government regulations and policies around land use and agriculture.
- Gathered a dataset of images and labeled our dataset: The dataset was stored on Zenodo platform https://doi.org/10.5281/zenodo.4628934.A dataset of 1117 images is considered adequate since YOLOv5 uses a transfer learning approach during the training.
- Export the dataset to YOLOv5.
- Trained YOLOv5 to recognize the objects in our dataset.
- Access the performance of YOLOv5 model.
- Execute test inference to see the model in action.

The four classes were balanced as shown in Fig. 2 that indicated the representation of each class. The Labels Correlogram in Fig. 3 indicates the annotation heatmap across all classes is a good representation of all datasets.

The number of training epochs to use can be a challenge when developing neural networks. The training dataset may become overfit if there are too many epochs, whereas there may be an underfit model if there are too few. Early stopping is a technique that enables you to specify an arbitrarily large number of training epochs and terminate training when the model's performance on a holdout validation dataset stops advancing [20]. During the training of the model, patience was set

Fig. 2 Class balance for poultry dataset for object detection

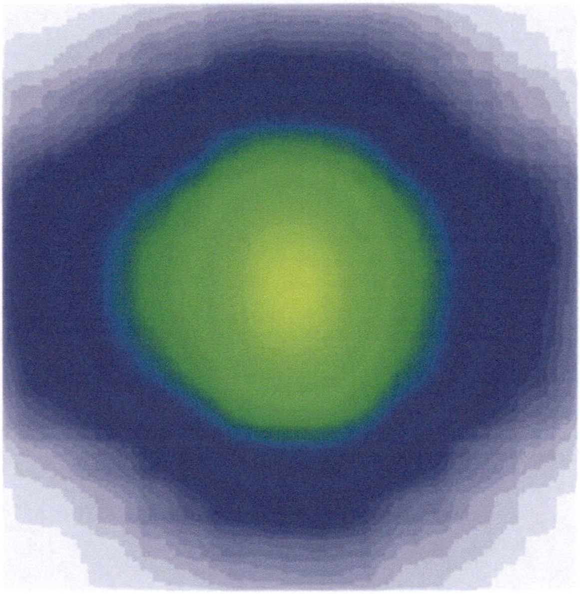

Fig. 3 Labels Correlogram

to 100 which means the training will stop if no improvement is observed in the last 100 epochs.

3 Results

The training metrics used for the object detection task in YOLOv5 were precision, recall, mean average precision with an IOU (Intersection Over Union) of 0.5 (mAP_0.5). The best results were observed at epoch 506. The training metrics with values are indicated in Fig. 4.

To evaluate the performance of object detection models like YOLO, the mean average precision (mAP), precision, and recall are used. The mAP compares the ground truth bounding box to the detected box and returns a score. The larger the score, the more accurate the model is in its predictions. The precision reflects how reliable the model is in classifying samples as Positive, it simply measures the percentage of your predictions that are correct. On the other hand, the recall metric tells us the extent our model can classify all the positive samples correctly as *Positive*.

From Fig. 4, the **mAP_0.5/ Step** graph which is the mean Average Precision (mAP) at IoU (Intersection over Union) threshold of 0.5 gave 87.25% confidence.

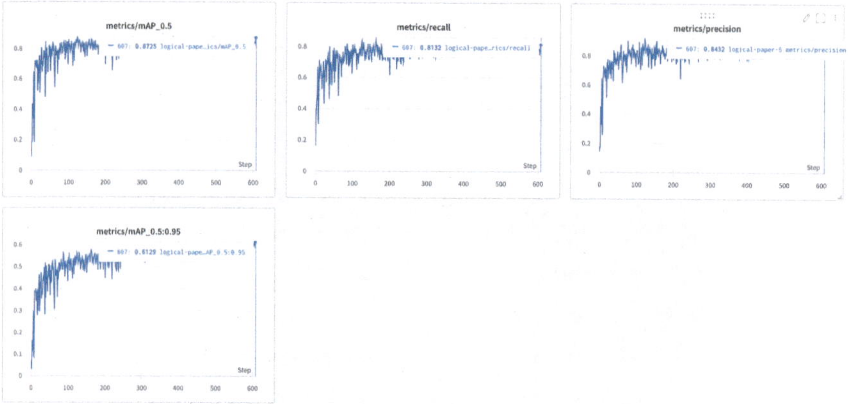

Fig. 4 Training metrics with values on YOLOv5

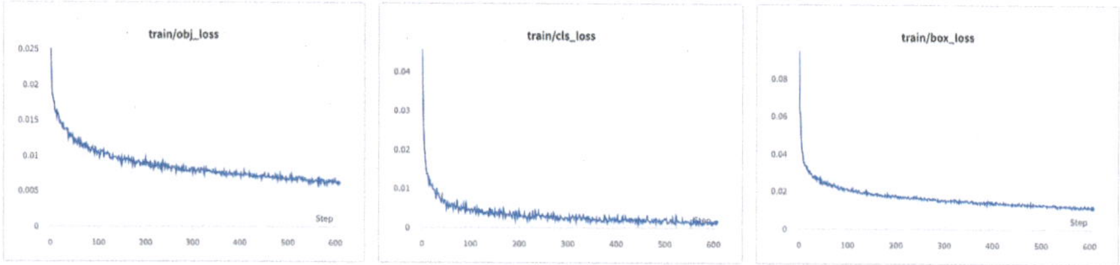

Fig. 5 Training loss

Precision/Step gave us a confidence of 84.32%. Also, the **recall/Step** graph gave us a confidence value of 81.32%.

Since early stopping was introduced during training, we observed the minimum loss that the model achieved in Fig. 5.

Figure 5 displays three different types of loss: classification loss, objectness loss, and box loss. The box loss measures how accurately the algorithm can locate an object's center and how perfectly the predicted bounding box completely covers an object. The probability that an object exists in a suggested region of interest is basically measured by objectness. If the objectivity is high, an object is probably present in the image window. How effectively the algorithm can accurately predict the correct class of a given object is indicated by the classification loss.

The box, classification, and objectness losses of the training data also showed a rapid drop until around epoch 20. The objectness loss attained a minimum value at around 0.0052, classification loss at 0.0025 and box loss at 0.01.

The confusion matrix for the four classes is indicated in Fig. 6.

The confusion matrix makes it easier to see if the model is confusing between the classes. You can get a better understanding of the classification model's successes and failures by calculating a confusion matrix. The expected row and predicted column for each class value holds the total number of accurate predictions for that

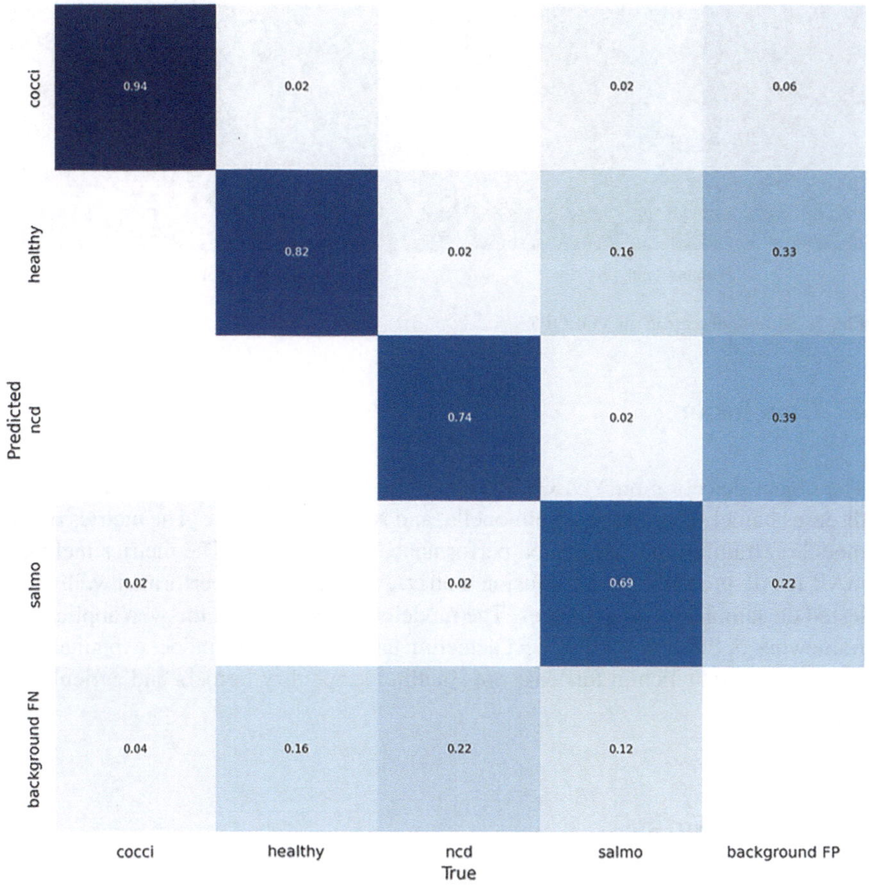

Fig. 6 Confusion matrix for four classes

class. In Fig. 6, Cocci, healthy, ncd, and salmo classes were correctly predicted with the confidence of 94%, 82%, 74% and 69%, respectively.

The model inference was done on Streamlit, a web application framework. Streamlit applications use Python scripts that execute from top to bottom. The script is re-executed every time a user opens a browser tab pointing to your application. Streamlit displays the script's output in real time in a browser as it runs.

The inference in Fig. 7 compares with the ground truth. On the testing sample images, the model detected cocci at 94% which implies a good confidence of accurate prediction of the cocci sample. Healthy sample was also accurately predicted with a confidence of 91%. We tried to capture a faint image of another healthy sample and it was also accurately predicted but with a low confidence level (32%).

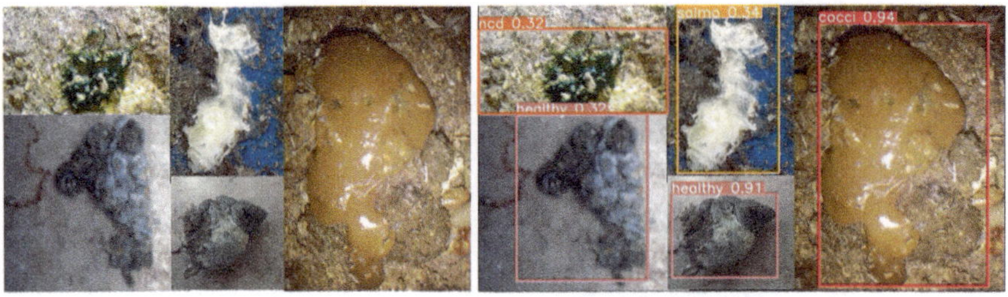

Ground Truth Predicted on YOLOv5

Fig. 7 Sample detection on YOLOv5

4 Conclusion

The object detection on YOLOv5 gave results for early detection of three poultry diseases namely Coccidiosis, Salmonella, and Newcastle disease. The metrics of the model on training indicated good performance of the model. The metrics included mAP, recall, precision, and confusion matrix. Also, the model performed well when tested on sample not-seen images. The model was deployed on the web application framework of Streamlit. The object detection task improves the model explainability for target users which in this case are smallholder poultry farmers and agricultural experts.

5 Future Works

From the findings in this work, we recommend deployment of the model on the mobile devices so that it can easily be accessible by the end users who in our case are poultry farmers. The mobile application has to be simple and user-friendly so that any stakeholder can easily interact with it. We also recommend future research on some other machine learning techniques like image segmentation to see if they can further improve the accuracy of the predictions.

References

1. Dorji, N., & Gyeltshen, T. (2012). Characterisation of family poultry production in Haa and Mongar districts of Bhutan. *Livestock Research for Rural Development, 24*(155), 2012.
2. Msami, H. (2007). *Poultry sector country review*. fao.org.
3. Moreki, J. C., Dikeme, R., & Poroga, B. (2010). The role of village poultry in food security and HIV/AIDS mitigation in Chobe District of Botswana. *Livestock Research for Rural Development, 22*(55), 1–7.

4. Kabatange, M. A., & Katule, A. M. (1989). Rural poultry production systems in Tanzania. In *Proceedings of an International Workshop on Rural Poultry Development in Africa* (pp. 13–16).
5. National Bureau of Statistics. (2007). Gender profile of smallholder rural agriculture population in Tanzania mainland, Dar es Salaam. www.kilimo.go.tz
6. FAO.org. (2013). Poultry development review. www.fao.org
7. Yongolo, M. G., Christensen, H., Handberg, K., Minga, U., & Olsen, J. E. (2011). On the origin and diversity of Newcastle disease virus in Tanzania. *Onderstepoort Journal of Veterinary Research, 78*(312), 1–8.
8. Knueppel, D., Cardona, C., Msoffe, P., Demment, M., & Kaiser, L. (2010). Impact of vaccination against chicken Newcastle disease on food intake and food security in rural households in Tanzania. *Food and Nutrition Bulletin, 31*(3), 436–445.
9. Msofe, P. L. M., Bunn, D., Muhairwa, A. P., Mtamob, M. M. A., Mtambo, H., Mwamhehe, A., Msago, M. M., & Cardona, C. J. (2010). Preparation for the prevention and control of highly pathogenic avian influenza in rural Tanzanian Village settings. *Avian Diseases, 54*(1), 768–771.
10. Swai, E. S., Sanka, P. N., & Kaaya, J. E. (2013). An investigation of the common causes of death in chicken around Arusha municipality area, Tanzania. *Livestock Research for Rural Development, 25*(11), 204.
11. Mitchell, A. (2015, August 1). The poultry site. [Online]. https://www.thepoultrysite.com/articles/identifying-poultry-diseases-in-your-flock
12. Mbelwa, H., Mbelwa, J., & Machuve, D. (2021). Deep convolutional neural network for chicken diseases detection. *International Journal of Advanced Computer Science and Applications, 12*(2).
13. Alarqoun, S., Baur, C., Achilles, F., Belagiannis, V., Demirci, S., & Navab, N. (2016). AggNet: Deep learning from crowds for mitosis detection in breast cancer histology images. *IEEE, 35*(5), 1313–1321.
14. Owomugisha, G., Quinn, J. A., Mwebaze, E., & Lwasa, J. (2014). Automated vision-based diagnosis of Banana bacterial wilt disease and black Sigatoka disease. In *1st International conference on the use of mobile ICT in Africa*, Stellenbosch.
15. Dasiopoulou, S., Mezaris, V., Kompatsiaris, I., Papastathis, V.-K., & Strintzis, M. G. (2005). Knowledge-assisted semantic video object detection. *IEEE Transactions on Circuits and Systems for Video Technology, 15*(10), 1210–1224.
16. Praharsha, V. (n.d.). OpenGenus. https://iq.opengenus.org/yolov4-model-architecture/
17. Bochkovskiy, A., Wang, C.-Y., & Liao, H.-Y. M. (2020). *YOLOv4: Optimal speed and accuracy of object detection*. Arxiv.
18. Jocher, G. (n.d.). pytorch.org. [Online]. https://pytorch.org/hub/ultralytics_yolov5/
19. Jocher, G. (2020). github.com. Ultralytics. [Online]. Retrieved 2022 from https://github.com/ultralytics/yolov5
20. Brownlee, J. (2018, December). machinelearningmastery.com. [Online]. https://machinelearningmastery.com/how-to-stop-training-deep-neural-networks-at-the-right-time-using-early-stopping/

Machine Learning Model for Predicting Construction Project Success in Tanzania

Astin Ntulo, Elizabeth Mkoba, Dina Machuve, and Sanket Mohan Pandhare

1 Introduction

The construction industry is one of the most important and vital sectors to any developing country because of its capital expenditure. This is because building infrastructures and facilities contribute to socio-economic development of the nation. Some of the construction projects include commercial buildings projects, residential buildings projects, infrastructure projects, healthy facilities, education facilities, and industrial plants construction projects. Construction project success is vital to ensure that projects contribute to the country's economic development and growth. According to [1], construction project success has been influenced by various factors such as commitment and involvement of both parties, schedules, experience of team members and availability, and senior management support.

Tanzania is among the developing countries with small and mega construction projects that are completed, in progress, and planned for the future. However, currently, the prediction and monitoring of a construction project's successful completion is done manually. Most of the projects use post-implementation evaluation method which has no impact on project being implemented successfully. Management behavior and experience are used as project control mechanisms [2].

With the current technological development, Artificial Intelligence (AI) is bringing new opportunities to the construction industry globally. Machine learning being a branch of artificial intelligence, has been playing a bigger role in "smart" construction [3]. By using machine learning algorithms, construction projects can be continuously monitored and even predict their successfulness in real time.

A. Ntulo (✉) · E. Mkoba · D. Machuve · S. M. Pandhare
Computational and Communication Science and Engineering (CoCSE), The Nelson Mandela African Institution of Science and Technology (NM-AIST), Arusha, Tanzania
e-mail: ntuloa@nm-aist.ac.tz

J. Marx Gómez et al. (eds.), *Artificial Intelligence Tools and Applications in Embedded and Mobile Systems*, Progress in IS,
https://doi.org/10.1007/978-3-031-56576-2_10

Random forest classification algorithm is one of the ensemble machine learning models, however, it has never been used before to predict construction project success in Tanzania. Thus, the objective of this study is to use random forest classification algorithm to predict construction project success in Tanzania. This is accomplished by close and continuous monitoring the critical successful factors for project success throughout its implementation. Critical success factors (CSFs) are defined as the characteristics, conditions, or variables which when maintained, sustained, and monitored properly provide a positive impact (benefits realization) for successful completion of a project [4]. A total of 41 factors were identified and used as inputs for training of the model.

This paper is organized as follows: Section 2 provides a literature review. The developed machine learning model for predicting construction project success in Tanzania is provided in Sect. 3. Section 4 presents the research methodology. Results and discussion are provided in Sect. 5. Finally, conclusion and recommendations for future research are presented in Sect. 6.

2 Literature Review

This section introduces literature reviews of different studies done on machine learning prediction models for construction project success and different machine learning approaches used in this study.

2.1 Machine Learning Models to Predict Construction Project Success

[1] proposed an artificial neural network-based prediction model for the performance of a construction project. The model was built on six fundamental elements that affect how well construction projects function. Different indices, such as cost, time, profit, and quality, were used to gauge these elements. The study chose the specifics of 40 projects as the dataset for the neural network. Involvement, commitment of both project partners (30.9%), timetable (25.4%), experience of project team (24.5%), and senior management support (14.3%) were found to have a high rate of influence on project performance (95.1%). The study is restricted to a dataset of 42 projects with a 96.1% prediction accuracy. The use of more data during training can increase the machine learning model's accuracy.

Furthermore, [2] proposed a project success prediction using an Evolutionary Gaussian Process Inference Model. The study uses historical dataset from the Continuous Project Performance (CAPP) database to train the model for the prediction of project success completion. CAPP is a tool developed by Construction Industry Institute (CII) that is based at The University of Texas at Austin in the USA and it contains large number of variables such as (1) actual owner expenditure, (2) actual design % complete, (3) construction % complete, (4) commitment for materials and equipment, (5) construction drawings, (6) personnel turnover, (7) schedule impact, and (8) incident rate, among others that are most influential in predicting project successfulness completion. The machine learning model development depends solely on the CAPP dataset that was developed by The University of Texas in the USA, using construction companies based in the USA. The project success factors may vary demographically due to different factors such as technological advancement hence the model developed using CAPP dataset can be limited to be used in other places.

According to [5], in her study she created a machine learning model for prediction and visualization of HIV index testing in northern Tanzania. The study includes three different machine learning algorithms (Random Forest, XGboost, and artificial neural network). The research results show that Random Forest performs better prediction compared to the other algorithms used in the study. The research was limited to the northern part of Tanzania only.

[6] studied how machine learning can be used to detect banana diseases in Arusha and Mbeya regions. In this study, the researcher used a machine learning technique called deep learning method using a convolution neural network (CNN). The model was developed and deployed in a mobile application where a farmer can take a picture of a banana leaf with a mobile phone, upload it to the model, and get results of the detected banana diseases. The model was limited only to banana leaves.

[3] did a study on machine learning and how it is applied in the construction industry. The study highlights the strengths and weaknesses of using machine learning in construction industry such as supervising sites, auto-detection of anomalies, and smart maintenance. Despite the fact that, the researcher points out that there is not enough dataset in the construction industry that can be used to train the machine learning model and that it impacts prediction accuracy. Thus, the study failed to account for other factors such as data preprocessing and parameter hyper tuning which have a great impact on machine learning model accuracy.

2.2 *Generative Adversarial Networks*

In this study, we adopted Generative Adversarial Networks (GANs) technique to expand the dataset. GANs are neural networks that were first introduced in a paper by Ian Goodfellow and his colleagues [7]. It is unsupervised learning that automatically learns the patterns of input data and produces synthetic data of the same behavior or characteristics as the original data. Basically, the generated fake data mimics the original input data. Using the GANs technique, our dataset expanded to 1082 from 97 inputs.

2.3 Random Forest Classification

Random forest is a type of supervised machine learning algorithm used globally for classification problem solving [8]. Random forest is basically an ensemble learning approach in which multiple classifiers provide solution to a complex problem. It works as a combination of tree classifications for predicting an outcome. Each tree makes a prediction, then the final output is the combination of each individual tree output.

2.4 Feature Selection

Feature selection is the automatic process of selecting the most consistent and relevant features that have a great contribution to building a machine learning model. This study opted for Mean Decrease in Impurity (MDI) technique for feature selection as it is widely preferred for a random forest model [9]. MDI automatically computes for feature importance by summing the decrease in node variance during data split process by assigning high importance to features with high MDI.

A total of 41 factors shown in Table 1 were identified through literature review and experts' opinions as critical success factors in construction project success. These factors were rated by construction industry experts to identify the importance of each factor's contribution to construction project success.

3 Research Methodology

The qualitative and quantitative research methods were employed in this research. Data were collected through survey questionnaires and interviews from construction industry experts and stakeholders. Experts and stakeholders include but are not limited to contractors, consultants, clients, engineers, and architects. The survey questionnaire included demographic data, understanding of the factors influencing construction projects' success and experience of the participants in the construction industry.

4 Model Development and Deployment

Dataset comprised 41 identified features as shown in Table 1, which are the factors for a construction project's success as identified by various studies and construction industry experts. These features were the inputs in the machine learning model while the output was successfulness of a construction project.

Table 1 Critical success factors for a construction project success

S/No	Project critical success factors
1	Completion time
2	Construction cost
3	Quality of a construction project
4	Communication between project stakeholders
5	Top management support
6	Project initial planning
7	Project requirements and objectives
8	Project reporting in project life cycle
9	Project manager competence
10	Project technical team experience and competence
11	Change management process
12	Project desired outcome
13	User/client's satisfaction
14	Project team satisfaction
15	Top management satisfaction
16	Client's financial capacity
17	Project team commitment
18	Project scope
19	Project risk management
20	Political environment
21	Competence in project management
22	Functional requirements
23	Monitoring and control
24	Effective anti-corruption policy
25	Project team size
26	Project team motivation
27	Project management methodologies
28	Environmental impact
29	Social-economic environment
30	Technological environment
31	Site limitation and location
32	Constructability
33	Formal dispute resolution process
34	Consultant's competence
35	Level of automation
36	Level of skilled labor required
37	Site inspections
38	Project vision and mission
39	Quality assurance
40	Real-time project auditing
41	Contract management and performance

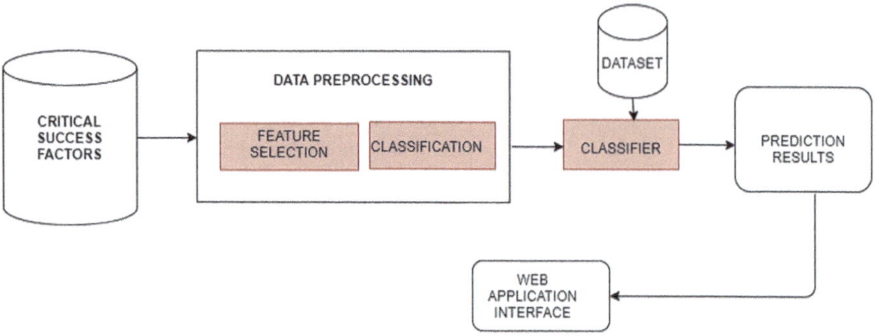

Fig. 1 Machine learning model web deployment block diagram

Data preprocessing and analysis were done using Python programming language, and Generative adversarial Networks (GANs) technique was used to expand the dataset. Seventy percent of the dataset were used for training and 30% of the dataset were used for testing. In feature selection, Mean Decrease in Impurity (MDI) techniques was used for feature importance in which, each variable factor examined its predictive impact [10]. The model was developed using random forest ensemble algorithms [11]. Streamlit framework was used for web-based model deployment. Streamlit is an open-source framework for developing web applications for machine learning [12]. It is compatible with a lot of Python libraries such as Pandas, Numpy, and Matplotlib. Figure 1 shows the block diagram of a machine learning model web deployment.

5 Results and Discussion

5.1 Results

The model was trained by random forest classifier, linear regression, decision tree classifier, and bagging classifier. Random forest classifier performed better with accuracy of 97.5% as shown in Table 2. With such accuracy, construction industry practitioners in Tanzania can use this model to accurately predict if the construction project will be successful or not.

Table 2 Prediction accuracy

S/No	Algorithm	% Prediction accuracy
1	Random forest classifier	97.5%
2	Bagging classifier	97.2%
3	Decision tree classifier	96.9%
4	Logic regression	96.3%

5.2 *Discussion*

Critical success factors influencing successful completion of a construction project in Tanzania as shown in Table 1 are unanimously agreed by several studies [1, 13]. This study shows that quality of the construction project, construction cost, and completion time are the most influencing factors for project success. These findings corroborate with different studies in the construction industry and it is sometimes referred to as the "Iron Triangle" of a successful construction project [14].

6 Conclusion

Ensemble machine learning techniques and random forest classification algorithms were used to develop the machine learning model for predicting construction project success in Tanzania. Random forest classifier had a better performance with an accuracy of 97.5% compared to other classifiers like logistic regression (96.3%) and bagging (97.2%). The model developed can help practitioners, decision makers, and policy makers to predict construction project success before project implementation.

Future research needs to examine more factors influencing project success in Tanzania as well as increase the number of datasets to measure the general performance.

References

1. Maya, R. (2021). Develop an artificial neural network (Ann) model to construction project performance in Sysria. *Journal of King Saud University—Engineering Sciences, 35*(6), 366–371.
2. Cheng, M. Y., Huang, C. C., & Roy, A. F. V. (2012). Predicting project success in construction using an evolutionary Gaussian process inference model. *Journal of Civil Engineering and Management, 19*(Sup1), S202–S211.
3. Xu, Y., Zhou, Y., Sekula, P., & Ding, L. (2021). Machine learning in construction: From shallow to deep learning. *Developments in the Built Environment, 6*, 100045.
4. Berssaneti, F. T. (2014). Identification of variables that impact project success in Brazil. *International Journal of Project Management, 33*(3), 638–649.
5. Chikusi, H. (2022). Machine learning model for prediction and visualization of HIV index testing in northern Tanzania. *International Journal of Advanced Computer Science and Applications, 13*(2).
6. Sanga, S. (2020). Mobile-based deep learning models for banana diseases detection. In *International Conference on Learning Representations*.
7. Goodfellow, I., et al. (2020). Generative adversarial networks. *Communications of the ACM, 63*(11), 139–144.
8. Breiman, L. (2001). *Random forests*. Kluwer Academic Publishers.
9. Xiao, et al. (2019). A Debiased MDI feature importance measure for random forests. In *33rd Conference on Neural Information Processing Systems (NeurIPS 2019)*, Vancouver, Canada.
10. Scornet, E. (2021). Trees, forests, and impurity-based variable importance in regression. *HAL*.

11. Dietterich, T. G. (2000). *Ensemble methods in machine learning*. Springer.
12. Singh, P. (2021). *Deploy machine learning models to production*. APRESS.
13. Cheng, M. Y., Wu, Y. W., & Wu, C. F. (2010). Project success prediction using an evolutionary support vector machine inference model. *Automation in Construction, 19*, 302–307.
14. Mohtar, S. (2013). The iron triangle as the Tripple constraints in Project Management. *Jurnal Teknik Sipil, 1*.

Improved Medical Imaging Transfer Learning through the Conflation of Domain Features

Raphael Wanjiku, Lawrence Nderu, and Michael Kimwele

1 Introduction

Deep learning has been used in many medical applications, from medical imaging classification in lung disease detection [1], pulmonary diseases detection [2], and image segmentation such as multi-modal imaging [3]. In medical imaging, transfer learning is widely used in facilitating deep learning. It has been used for skin cancer detection and classification [4], Alzheimer's disease detection [5], chest X-rays classification [6, 7], diabetic foot ulcers classification [8], and histopathology image classification, among others [9].

Transfer learning has become the more straightforward route in using deep learning in medical imaging classification and segmentation. Due to the difference in pre-trained model training, pre-trained models are fine-tuned reasonably on the bottom and top layers in other applications [10]. Transfer learning involves reusing previously learnt knowledge of one domain in another domain. For example, one can reuse the knowledge of driving a car to drive a lorry. However, transfer learning works in closely related tasks.

ImageNet pre-trained models have become the go-to approach in adapting pre-trained models with their pre-trained weights, fine-tuning them to adapt to the given tasks [6]. Some of the adopted architectures include the ResNet on chest X-rays and retinal fundus [11], skin cancer detection [4], and IVF procedures [12]. However, there is always a question of the helpfulness of feature reuse for medical images due to variations in textures in pathology (retinal fundus, patches in pneumonia) [6]. This variation of features problem can be addressed with features conflation in

R. Wanjiku (✉) · L. Nderu · M. Kimwele
Jomo Kenyatta University of Agriculture and Technology, Nairobi, Kenya

J. Marx Gómez et al. (eds.), *Artificial Intelligence Tools and Applications in Embedded and Mobile Systems*, Progress in IS,
https://doi.org/10.1007/978-3-031-56576-2_11

transfer learning in both domains to improve target data adaptation, thereby assisting radiology practitioners in bridging this feature variation gap.

Another contentious issue in transfer learning is understanding the success factors of transfer learning and the elements responsible. Over many studies, researchers have noted that models' lower layers [13] and feature reuse, especially low-level features, play a critical role in success transfer [10]. The importance of early pre-trained model layers is observed to contribute to the reuse of these local features, which require vast amounts of data to learn [14]. Therefore, this work addresses poor target domain data adaptation using the conflation of textural features between the source and target domain in transfer learning tasks.

2 Literature Review

The following literature presents the various uses of transfer learning in medical imaging and textural methods in feature extraction.

In a study by Rajpurkar et al. [15], the researchers used X-ray images to detect pneumonia using a CNN model dubbed CheXNet trained on the ChestXray14 dataset with 14 disease types [15]. Minaee et al. [16] use transfer learning on COVID-19 radiology images to detect infected patients and aid decision-making for proper care administration. The researchers used ResNet18, ResNet50, DenseNet-121, and SqueezeNet to identify the disease using heat maps to show the lung sections infected [16]. The proposed approach is different through the conflation of features and adds a MobileNetV2, further showing its applicability to low-resource environments. However, the researchers in COVID-19 detection used more pre-trained models, giving more dataset generalization.

In a further study by Peng et al. [10], the researchers report that low-level features play a critical role in the transfer process when evaluating medical image classification. Raghu et al. [6] attribute low-level statistics as one of the elements that enable successful transfer in medical tasks [6]. They note that when there is a slight difference between the source and target, there is more success. The evaluation of low-level is similar to the proposed approach but does not consider the conflation of these features.

In a study by Henna and Raji [17], the researchers use Contrast limited adaptive histogram equalization (CLAHE): a data augmentation approach to augment X-ray images to eliminate transfer learning bias for COVID-19 detection. To fine-tune the dataset, they use various models, including AlexNet, ResNet, and VGG16 [17]. This approach is different from the proposed approach through the use of augmentation. The augmentation process can still be applicable in this proposed model only after identifying the quality samples that improve performance.

Other studies involving the use of medical imaging include skin disease diagnosis. In a study by Bevan and Atapour-Abarghouei [18], the researchers propose using a single model-based strategy to classify skin lesions on small and imbalanced datasets to remove bias and spurious variations from the melanoma classification pipeline. The strategy uses Dropout and DropBlock regularization methods to reduce overfitting and introduces the Modified Rand Augment augmentation strategy. They use the ISIC 2017 dataset. This proposed approach uses the ISIC 2016 dataset, which is smaller and well-suited for transfer learning tasks. The researchers' model uses augmentation to increase the data samples and regularization techniques which have not been used in the proposed approach. However, the benefits of regularization and augmentation could only improve the performance of the proposed approach, as evident in the existing literature [17].

Among the studies done in bridging the domain, data adaptation includes [19, 20] and the use of the Maximum-likelihood principle by Savchenko [21], which uses D_{KL} but does not use conflation of features in identifying matching features.

3 Methodology

This work proposes the use of conflation of textural probability distribution features in the medical imaging datasets and those of the pre-trained model training dataset. The conflated features are single-dimensional vectors (probability distributions); these distributions are compared using the Kullback-divergence measure (D_{KL}) and ranked to pick the images whose divergence is lower than the mean of the sum measures of all the samples. Kullback-Leibler divergence is selected due to its reasonable time and memory complexity.

3.1 Probability Distributions Conflation

The conflation of probability distributions (&) is a concept introduced by Hill [22] and refers to merging any two probability distributions. Given a set of distributions $\{Q_1, \ldots, Q_n\}$, the conflation distribution X is expressed in Eq. 1:

$$X = \&\,(Q_1, \ldots, Q_n) \tag{1}$$

X can be further expressed using the probability density functions and the data samples for discrete and continuous probability distributions, as shown in Eqs. 2 and 3, respectively:

$$\&\,(Q_1, \ldots, Q_n) = \frac{f_1(k) f_2(k), \ldots, f_n(k)}{\sum_y f_1(l) f_2(l), \ldots, f_n(l)} \tag{2}$$

$$\&\,(P_1, \ldots, P_n) = \frac{f_1(k) f_2(k), \ldots, f_n(k)}{\int_{-\infty}^{\infty} f_1(l) f_2(l), \ldots, f_n(l)} \tag{3}$$

where $f_1, \ldots, f_n$ is the probability density function of the probability distributions $P_1, \ldots, P_n$, while k and l are the data samples.

3.2 The Conflation of Feature Vectors

Consider an image I with a set of extracted low-textural features and convert the features into probability distributions using the softmax function. With probability densities $\left\{p\left(I_1^f\right), \ldots, p\left(I_n^f\right)\right\}$ for image features 1 to n, we can conflate the features of I such that:

$$\&\left(I_1^f, \ldots, I_n^f\right) = \frac{\prod_f p\left(I_{1n}^f\right)}{\int \prod_f p\left(I_{1n}^f\right)} \tag{4}$$

where p is the probability distribution density, f is an image feature, and n is the number of features in the image.

We can iterate the conflation process for each image in both the source and the target domains generating merged probability distributions for all the images in both domains. After that, we compare each target image to the source images using the Kullback-Leibler divergence, as shown in Eq. 5.

$$D_{KL}\left(P\left(I_{T1}^f\right) \parallel P\left(I_{S1}^f\right)\right) = \int_{-\infty}^{+\infty} \&P\left(I_{T1}^f\right)(x) \log \frac{\&P\left(I_{T1}^f\right)(x)}{\&P\left(I_{S1}^f\right)(x)} dx \tag{5}$$

where $P\left(I_{T1}^f\right)$ is the probability distribution of features of an image 1 in the target domain, $P\left(I_{S1}^f\right)$ is the probability distribution of features of an image 1 in the source domain, T refers to the target domain dataset samples, and S is the source domain dataset sample.

A mean $\overline{x}$ of the D_{KL} values is then calculated, and any target image whose mean is below the average is selected as a candidate for use in the transfer learning process.

Some of the studies involving the conflation of features include dimensionality reduction in large datasets by Mitra et al. [23], Rahman et al. [24], Gan et al. [25], and many studies involving Geographic information systems (GIS) [26].

Table 1 Datasets distribution

Dataset	Training	Validation	Classes
COVID-19 radiography dataset	14,815	6350	4
ChestX-ray8	3200	800	4
ISIC melanoma	900	379	2

3.3 Experiments

Datasets

The experiments conducted in this work utilize three publicly available datasets, as outlined in Table 1.

The data samples for the chest X-ray are a subset of the larger dataset, which contains over 108,000 chest images. Also, only four thorax disease categories (Atelectasis, Infiltration, Mass, and Nodule) are picked from the dataset from the 14 thorax disease categories.

Pre-Trained Models

The experiments are conducted on two pre-trained model architectures: MobileNetV2 and DenseNet121. The MobileNetV2 was selected to determine the performance on the resource-limited devices, while DenseNet121 was selected as the recommended model architecture for working with medical images [27].

Experimental Settings

The experiments are performed on the PaperSpace platform (Quadro P5000 8CPU 30GB RAM GPU) using Tensor flow 2.4.1. The Stochastic Gradient descent optimizer is used with a learning rate of 0.001. The softmax function is the classification function in the 100 epochs training. A Dropout layer with a rate of 0.5 and a Batch normalization layer is added to regularize the models.

Experimental Methods

Features extraction from the dataset images is acquired using the pre-trained models' initial convolutional layers as opposed to the traditional feature extraction methods. This extraction is done to simulate the standard processing of an image when using a pre-trained model without prior feature extraction.

This study uses two feature analysis methods: Linear Binary Pattern (LBP) and Grey-Level Co-Occurrence Matrix (GLCM). The LBP compares a pixel's value to its neighbors with a block of 3×3 or more, returning a feature vector of the gray levels. The feature vector is then converted to a decimal value that encodes the extracted textural features of an image. LBP is preferred due to its powerful monotonic illumination of variations and simplicity [28].

The GLCM is a textural feature descriptor that returns a matrix of the different combinations of gray levels in an image. This study uses three properties to aid the domain data adaptation description: energy, correlation, and homogeneity. The three properties are selected for this study due to their interdependence. The GLCM is used primarily due to its good feature description that reflects the spatial shape attributes of the gray's interval, direction, and amplitude differences [28].

4 Results and Discussion

The results from the experiments measure the performance of the samples selected using accuracy, precision, and recall by comparing them to the two selected architectures.

4.1 Performance Measures Comparison on MobileNetV2

From the results shown in Tables 2, 3, and 4, the lower values of D_{KL} give better accuracy, precision, and recall. The values are presented in percentages, where ABV represents values above the mean D_{KL} and BLW represents values below the mean D_{KL}.

Table 2 Accuracy comparison on MobileNetV2

	GLCM							
	Correlation		Energy		Homogeneity		LBP	
Dataset	*ABV*	*BLW*	*ABV*	*BLW*	*ABV*	*BLW*	*ABV*	*BLW*
COVID-19	**88.90**	**94.38**	91.29	92.87	89.88	94.18	**91.56**	**94.50**
ChestX-ray8	66.91	73.38	70.65	76.12	72.51	**77.54**	73.49	73.71
Melanoma	88.75	**92.88**	89.94	91.58	90.56	91.93	88.84	**93.59**

Table 3 Precision comparison on MobileNetV2

	GLCM							
	Correlation		Energy		Homogeneity		LBP	
Dataset	*ABV*	*BLW*	*ABV*	*BLW*	*ABV*	*BLW*	*ABV*	*BLW*
COVID-19	**90.57**	**95.13**	92.10	93.55	91.43	94.60	**93.54**	**94.83**
ChestX-ray8	72.81	78.98	75.31	80.80	78.67	**81.29**	79.16	81.05
Melanoma	88.75	**92.88**	89.94	91.58	90.56	91.93	88.84	**93.59**

Table 4 Recall comparison on MobileNetV2

Dataset	GLCM						LBP	
	Correlation		Energy		Homogeneity			
	ABV	*BLW*	*ABV*	*BLW*	*ABV*	*BLW*	*ABV*	*BLW*
COVID-19	86.05	**93.46**	90.01	91.28	87.89	93.33	89.03	92.15
ChestX-ray8	52.57	62.05	57.01	66.73	63.9	**70.02**	62.33	63.93
Melanoma	88.75	**92.88**	89.94	91.58	90.56	91.93	88.84	**93.59**

In Table 2, LBP performs better than the GLCM properties, with correlation performing well and energy giving the least values.

A similar trend is seen in Table 3 where LBP gives the two highest precision scores in the MobileNetV2, while correlation and homogeneity give good scores among the GLCM properties.

It can also be noted that recall values are lower in all three datasets, as seen in Table 4. However, the lower values of the three measures of DKL give the best results for the three datasets. Figure 1 shows the accuracy plots with values below the mean D_{KL} giving the best plots. As seen in the chart, the datasets with the lower D_{KL} values are seen to give better learning through the epochs and higher accuracy values.

4.2 Performance Comparison of Selection Methods on DenseNet121

Tables 5, 6, and 7 show that D_{KL} values below the mean for the DenseNet121 still show better performances, validating the results seen in the MobileNetV2 model.

LBP performs better than GLCM properties, while energy gives the highest accuracy value among the GLCM properties. Correlation gives the least accuracy values for the three datasets, as evident in Table 5.

LBP still gives the highest precision values among the three datasets, as seen in Table 6. Similarly, the correlation still yields lower values than the other GLCM properties, with homogeneity giving the highest GLCM values.

As seen in Table 7, LBP still gives the highest recall values, further highlighting the dominance of LBP over the GLCM properties. Energy and homogeneity still give the best values among the GLCM properties.

Among the GLCM properties, homogeneity and energy give better results in the DenseNetV121 model, as seen in Tables 5, 6, and 7. Homogeneity does well, as highlighted in the MobileNetV2 model results.

In Fig. 2, the samples with lower D_{KL} are seen to give good results as the number of epochs increases during the model training, a case also observed with the test data samples.

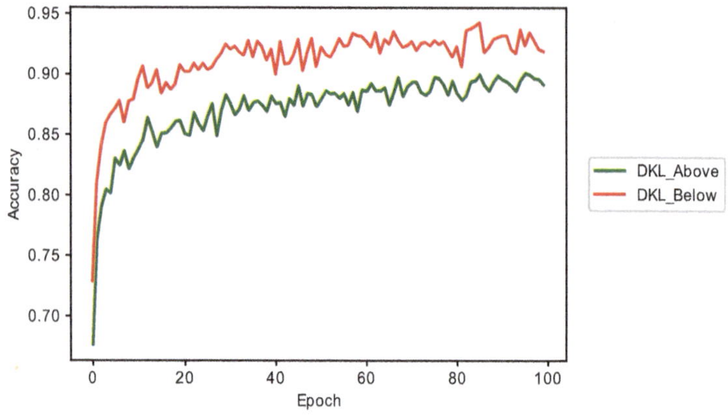

(a) LBP covid Accuracy

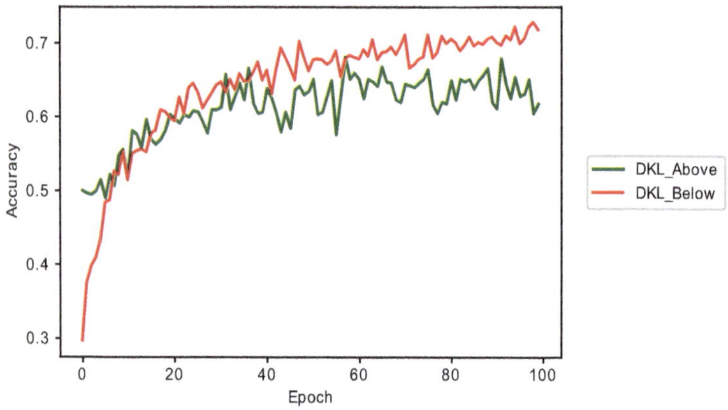

(b) Correlation ChestX-ray Accuracy

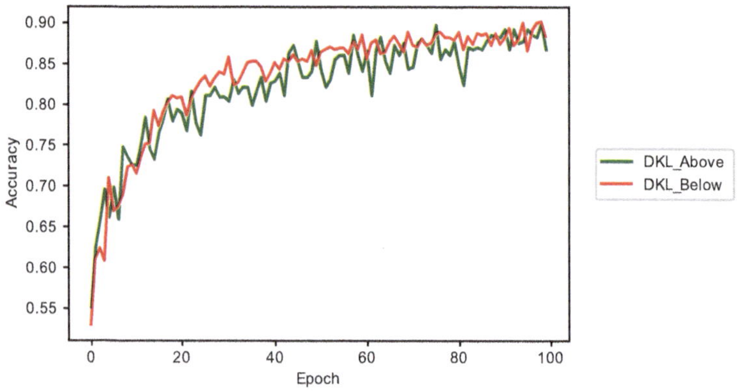

(c) Hemogeneity Melanoma Accuracy

Fig. 1 MobileNetV2 Accuracy performance plots on the three datasets

Table 5 Accuracy comparison on DenseNet121

	GLCM							
	Correlation		Energy		Homogeneity		LBP	
Dataset	*ABV*	*BLW*	*ABV*	*BLW*	*ABV*	*BLW*	*ABV*	*BLW*
COVID-19	90.35	91.02	90.10	91.45	90.32	**93.46**	90.18	**93.07**
ChestX-ray8	71.77	76.26	73.85	**77.02**	73.35	76.92	72.46	**85.63**
Melanoma	87.40	89.72	88.46	**93.58**	87.57	89.50	87.05	**93.04**

Table 6 Precision comparison on DenseNet121

	GLCM							
	Correlation		Energy		Homogeneity		LBP	
Dataset	*ABV*	*BLW*	*ABV*	*BLW*	*ABV*	*BLW*	*ABV*	*BLW*
COVID-19	91.82	92.26	92.00	92.79	91.30	**93.75**	91.40	**93.40**
ChestX-ray8	78.89	79.81	78.74	81.20	79.91	**82.01**	78.59	**87.65**
Melanoma	87.40	89.72	88.46	**93.58**	87.57	89.50	87.05	**93.04**

Table 7 Recall comparison on DenseNet121

	GLCM							
	Correlation		Energy		Homogeneity		LBP	
Dataset	*ABV*	*BLW*	*ABV*	*BLW*	*ABV*	*BLW*	*ABV*	*BLW*
COVID-19	88.34	89.50	88.28	89.77	88.76	**92.49**	89.21	**90.17**
ChestX-ray8	57.88	69.46	64.54	**69.53**	64.89	67.42	62.29	**78.70**
Melanoma	87.40	89.72	88.46	**93.58**	87.57	89.50	87.05	**93.04**

From the above results, GLCM properties could perform relatively well to the LBP due to their less discriminating features in the grey levels [28]. The overall performance of homogeneity shows that the uniformity and simplicity of the textural features give good feature adaptation hence better values. Homogeneity also increases the density of the edges and reduces the distances between the textural patches. This result is also actual for the correlation that increases sensitivity leading to good overall accuracy values in the MobileNetV2. The LBP performance is attributed to its invariance in illumination variations of the textural patches in the image features.

5 Conclusion

This work has introduced the conflation of textural features in three medical imaging datasets on two pre-trained models. The textural feature analysis has used the GLCM and LBP methods, and it can be seen from the results in Tables 3, 4, 5, 6, and 7 that textural features affect a pre-trained model's performance as a result of better domain data adaptation or lack of it. As noted in the results, selecting

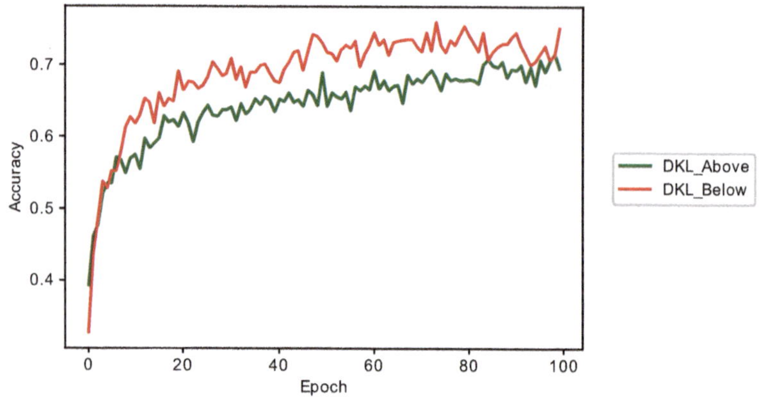

(a) Energy chest X-ray Accuracy

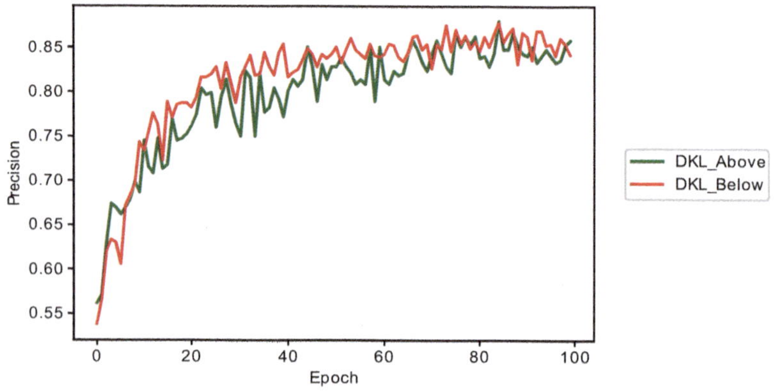

(b) Homogeneity Melanoma Precision

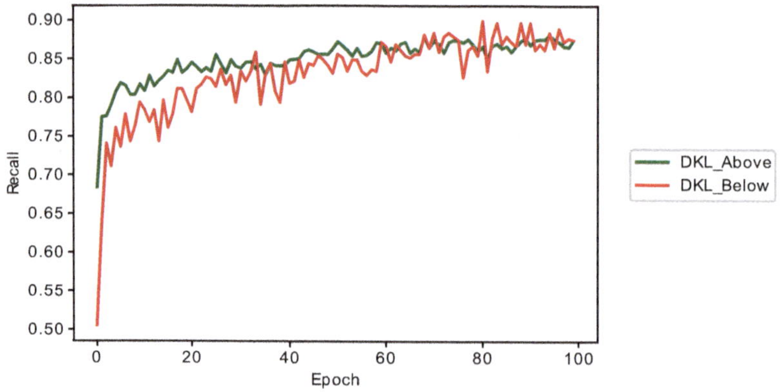

(c) LBP Covid Recall

Fig. 2 DenseNetV121 performance plots on the three datasets features with lower divergence values can improve the performance adaptation of transfer learning. These effects coincide with existing literature using divergence in evaluating information loss. This approach can be used in medical imaging and other scenarios requiring image classification using transfer learning. However, it is limited to working with image data. Further work can be performed on other pre-trained models and datasets to validate the results. Additionally, other low-level features can be explored to determine their effects on domain data adaptation and conflation of other features that do not have to be image related, such as signals in recurrent neural networks.

Data Availability Statement All the datasets are available via ReadMe in the code repository, and the source code can be found at https://github.com/geekhack/ICTA-EMOS_Medical-Imaging-Conflated-Features.

References

1. Roth, H., Lu, L., Farag, A., Shin, H.-C., Liu, J., Turkbey, E. B., & Summers, R. M. (2015). Deeporgan: Multi-level deep convolutional networks for automated pancreas segmentation. In *MICCAI* (pp. 556–564). Springer.
2. Wang, X., Peng, Y., Lu, L., Lu, Z., Bagheri, M., & Summers, R. M. (2017). ChestX-Ray8: Hospital-scale chest X-ray database and benchmarks on weakly-supervised classification and localisation of common thorax diseases. In *2017 IEEE Conference on Computer Vision and Pattern Recognition (CVPR)* (pp. 3462–3471).
3. Russakovsky, O., Deng, J., Su, H., Krause, J., Satheesh, S., Ma, S., Huang, Z., Karpathy, A., Khosla, A., Bernstein, M., et al. (2015). Imagenet large scale visual recognition challenge. *International Journal of Computer Vision, 115*(3), 211–252.
4. Esteva, A., Kuprel, B., Novoa, R. A., Ko, J., Swetter, S. M., Blau, H. M., & Thrun, S. (2017). Dermatologist-level classification of skin cancer with deep neural networks. *Nature, 542*(7639), 115–118. https://www.nature.com/articles/nature21056
5. Ding, Y., Sohn, J. H., Kawczynski, M. G., Trivedi, H., Harnish, R., Jenkins, N. W., Lituiev, D., Copeland, T. P., Aboian, M. S., Mari Aparici, C., Behr, S. C., Flavell, R. R., Huang, S. Y., Zalocusky, K. A., Nardo, L., Seo, Y., Hawkins, R. A., Hernandez Pampaloni, M., Hadley, D., et al. (2019). A deep learning model to predict a diagnosis of Alzheimer disease by using ^{18}F-FDG PET of the brain. *Radiology, 290*(2), 456–464. https://doi.org/10.1148/radiol.2018180958. Epub 2018 Nov 6. PMID: 30398430; PMCID: PMC6358051.
6. Raghu, M., Zhang, C., Kleinberg, J., & Bengio, S. (2019). Transfusion: Understanding transfer learning for medical imaging. In *Proceedings of the 33rd International Conference on Neural Information Processing Systems*.
7. Irvin, J., Rajpurkar, P., Ko, M., Yu, Y., Ciurea-Ilcus, S., Chute, C., Marklund, H., Haghgoo, B., Ball, R., Shpanskaya, K., Seekins, J., Mong, D. A., Halabi, S. S., Sandberg, J. K., Jones, R., Larson, D. B., Langlotz, C. P., Patel, B. N., Lungren, M. P., et al. (2019). CheXpert: A large chest 11 radiograph dataset with uncertainty labels and expert comparison. AAAI'19/IAAI'19/EAAI'19.
8. Bloch, L., Brungel, R., & Friedrich, C. (2021). *Boosting EfficientNets ensemble performance via pseudo-labels and synthetic images by pix2pixHD for infection and Ischaemia classification in diabetic foot ulcers*. DFUC@MICCAI.
9. Kim, H. E., Cosa-Linan, A., Santhanam, N., Jannesari, M., Maros, M. E., & Ganslandt, T. (2022). Transfer learning for medical image classification: A literature review. *BMC Medical Imaging, 22*(1), 69. https://doi.org/10.1186/s12880-022-00793-7. PMID: 35418051; PMCID: PMC9007400.

10. Peng, L., Liang, H., Li, T., & Sun, J. (2021). Rethink transfer learning in medical image classification. *Journal of Visual Communication and Image Representation, 78*, 2021.
11. Abràmo, M. D., Lou, Y., Erginay, A., Clarida, W., Amelon, R., Folk, J. C., & Niemeijer, M. (2016). Improved automated detection of diabetic retinopathy on a publicly 9 available dataset through integration of deep learning. *Investigative Ophthalmology & Visual Science, 57*(13), 5200–5206.
12. Khosravi, P., Kazemi, E., Zhan, Q., et al. (2019). Deep learning enables robust assessment and selection of human blastocysts after in vitro fertilisation. *Npj Digital Medicine, 2*, 21. https://doi.org/10.1038/s41746-019-0096-y
13. Neyshabur, B., Sedghi, H., & Zhang, C. (2020). What is being transferred in transfer learning? Paper presented at: NIPS'20: Proceedings of the 34th International Conference on Neural Information Processing Systems, 2020; Vancouver, Canada.
14. Matsoukas, C., Haslum, J. F., Sorkhei, M., Söderberg, M., & Smith, K. (2022). Proceedings of the IEEE/CVF Conference on Computer Vision and Pattern Recognition (CVPR) (pp. 9225–9234).
15. Rajpurkar, P., Irvin, J., Ball, R. L., Zhu, K., Yang, B., Mehta, H., Duan, T., Ding, D., Bagul, A., Langlotz, C. P., Patel, B. N., Yeom, K. W., Shpanskaya, K., Blankenberg, F. G., Seekins, J., Amrhein, T. J., Mong, D. A., Halabi, S. S., Zucker, E. J., et al. (2018). Deep learning for chest radiograph diagnosis: A retrospective comparison of the CheXNeXt algorithm to practicing radiologists. *PLoS Medicine, 15*(11), e1002686.
16. Minaee, S., Kafieh, R., Sonka, M., Yazdani, S., & Soufi, G. J. (2020). Deep-COVID: Predicting COVID-19 from chest X-ray images using deep transfer learning. *Medical Image Analysis, 65*, 101794–101794.
17. Henna, S., & Reji, A. P. (2021). A data augmented approach to transfer learning for Covid-19 detection. ArXiv, abs/2108.02870.
18. Bevan P. J., & Atapour-Abarghouei, A. (2021). Skin deep unlearning: Artefact and instrument debiasing in the context of melanoma classification. ArXiv.https://10.48550/arXiv.2109.09818
19. Ge, W., & Yu, Y.. (2017). Borrowing treasures from the wealthy: Deep transfer learning through selective joint fine-tuning. Paper presented at: 2017 IEEE Conference on Computer Vision and Pattern Recognition (CVPR); Honolulu.
20. Luo, Z., Zou, Y., Hoffman, J., & Fei-Fei, L. (2017). Label efficient learning of transferable representations across domains and tasks. Paper presented at: NIPS'17: Proceedings of the 31st International Conference on Neural Information Processing Systems; Long Beach, California.
21. Savchenko, A. V. (2017). Deep convolutional neural networks and maximum-likelihood principle in approximate nearest neighbor search. Paper presented at: Iberian Conference on Pattern Recognition and Image Analysis.
22. Hill, T. P. (2011). Conflations of probability distributions. *Transactions of the American Mathematical Society, 363*(6), 3351–3372.
23. Mitra, S., Saha, S., & Hasanuzzaman, M. (2020). Multi-view clustering for multi-omics data using unified embedding. *Scientific Reports, 10*(1), 13654.
24. Rahman, M. M., Voyles, R., & Wachs, X. (2021). Sequential prediction with logic constraints for surgical robotic activity recognition. Paper presented at: 2021 30th IEEE international conference on Robot & Human Interactive Communication (RO-MAN); Vancouver.
25. Gan, Z., Singh, P. D., Joshi, A., et al. (2017). Character-level deep conflation for business data analytics. Paper presented at: 2017 IEEE International Conference on Acoustics, Speech and Signal Processing (ICASSP); New Orleans.
26. Walter, V., & Fritsch, D. (2010). Matching spatial data sets: A statistical approach. *International Journal of Geographic Information Science, 13*(5), 445–473.
27. Hasan, N., Bao, Y., Shawon, A., & Huang, Y. (2021). DenseNet convolutional neural networks application for predicting COVID-19 using CT image. *SN Computer Science, 2*(5), 389. https://doi.org/10.1007/s42979-021-00782-7
28. Wang, S., Han, K., & Jin, J. (2019). Review of image low-level feature extraction methods for content-based image retrieval. *Sensor Review, 39*(6), 783–809.

Determining Emotion Intensities from Audio Data Using a Convolutional Neural Network

Simon Kipyatich Kiptoo, Kennedy Ogada, and Tobias Mwalili

1 Introduction

Human beings express their emotions in three major ways; speech, facial expressions, and textual messages [1]. In speech, the sounds produced by the person vary greatly with the intensity of the emotions being experienced at the time. Emotion detection has been done extensively on facial, textual, and speech using different methodologies based on the current technology. The intensities exhibited by these emotions have also been determined to some extent. However, it is observed that while it is easier to determine the emotion intensities in facial expressions and textual messages, emotion intensities in speech prove quite challenging due to several factors. Unlike facial expressions and textual messages, speech is very dynamic and the change in tonality depicts a mood swing [2]. These mood swings affect the pitch at which the sound is produced. This in turn affects the frequency at which a certain emotion is felt. Most emotions lie almost in the same frequency spectrum as indicated in Plutchik's wheel of emotions [3]. Emotions belong to two broad classes; discreet or continuous, time domain, or frequency domain [4]. To correctly map out emotion intensities discreet features versus continuous features have to be determined. Conversion of analog signals to digital signals is vital in correctly determining the frequency range of each emotion intensity. The dynamic changes in frequency make it more challenging for emotion intensity to be determined in order to pinpoint an emotion correctly. The tonal variations reflect the intensity of the emotional state of the speaker at the time of production [5].

Emotions change as conversations progress and coupled with sarcasm in the speaker's dialogue, the dynamicity of emotions becomes challenging to detect.

S. K. Kiptoo (✉) · K. Ogada · T. Mwalili
Jomo Kenyatta University of Agriculture and Technology, Nairobi, Kenya
e-mail: kodhiambo@scit.jkuat.ac.ke; tmwalili@jkuat.ac.ke

J. Marx Gómez et al. (eds.), *Artificial Intelligence Tools and Applications in Embedded and Mobile Systems*, Progress in IS,
https://doi.org/10.1007/978-3-031-56576-2_12

Due to this dynamicity, frequencies tend to overlap thus providing a challenge to correctly distinguish between the boundaries of each emotion. Fear and terror might be taken as one emotional state, but one is more intense than another. Similarly, Anger and Rage are considered to be one emotional state, however they differ in intensity. The most difficult part in differentiating these states is knowing the boundaries [6]. The best approach to emotion recognition from audio files is by using machine learning methodologies such as Neural Networks [7] that build models for classification of emotions based on feature extraction from audio dataset. Audio data contain two types of features which can be extracted and classified for emotion detection; Lexical and acoustic features. Lexical features are obtained from the spoken word or phonemes while the acoustic features represent the observed sound features during recording such as frequency, waveform, and amplitude [8]. The extracted features are then loaded into machine learning models as input.

In this paper, we propose to use a Convolutional Neural Network (CNN) to determine the intensities exhibited by emotions from an audio dataset. First, we intend to detect the emotions from MFCC extracted features, and then map each emotion to one of the intensities; strong or normal. The audio dataset will be subjected to a feature extraction process using an MFCC algorithm to obtain the coefficients necessary to classify the emotions. A Logistic Regression classifier will be used in the classification of the extracted features. MFCC coefficients are ideal due to their close semblance with the human ear perception mechanism. These features will be introduced to a model built from a Convolutional Neural Network to determine the intensities exhibited by each emotion.

2 Related Works

Several studies relating to emotion intensity determination have been done by various researchers using different methodologies. It is, however, observed that while successful experiments have been achieved by several researchers, these experiments mainly center around databases containing datasets from facial images and textual datasets. Emotion intensity determination from audio data has proved to be challenging and elusive. Little or nothing has been done on emotion intensity determination from audio data.

Megumi & Minoru [9] presented three models to analyze the level of emotional intensity of speech and estimate emotion intensity, do individual, parallel, and serial estimation. For individual estimation, each emotion and its intensity are independently considered and are categorized into one of the ten categories. Parallel estimation model creates separate estimators for emotion and intensity, then it determines the emotional intensity by integrating the outcome from the two estimators into one. For serial estimation model, the estimation of emotions is first done and put into one of the four emotional classes before classifying them into one of the appropriate classes. They used a chat corpus from an online gaming speech in conjunction with EmoDB to train and test a deep learning algorithm.

Each speech was recorded at a duration of 218 ms–7393 ms and was down sampled from 44.1 kHz to 16 kHz to match the EmoDB. From the overall outputs, two feature vectors were produced. Two classifiers; Support Vector Machine (SVM) and Random Forest were used to classify the utterance level from the two vector features. Of the three models, individual estimation and parallel estimation models each posted an average of 52.6% accuracy, while serial estimation posted an average accuracy of 52.3% compared to previous models.

Selvarajah et al. [10] proposed a deep hybrid Convolutional Neural Networks approach in estimating emotion intensity from video frames, containing a single CNN framework. For evaluation, they used two datasets from two different corpus; the UNBC-McMaster Shoulder Pain Expression Archive and Extended Cohn Kanade (CK+) datasets. UNBC contains 129 (63 males and 66 females), spontaneously recorded videos of subjects with shoulder pain. The CK+ contains 593 posed sequences. They considered the dominant Action Units (AUs) only in constructing various intensity levels. The AUs are actually the facial action unit-based emotions. They used a 5-point scale to indicate the intensity of each emotion; 1 representing low intensity and 5 representing high intensity. The intensities of five basic emotions; happy, angry, disgust, fear, and surprise were determined. The framework achieved a 91.45% accuracy classification on CK+ corpus and a 92.34% accuracy classification on UNBC corpus on the binary classification. According to the experiment, this outcome posted a lesser accuracy from their approach by 8% and 7%, respectively. For the five-class intensity estimation, the framework achieved a 74.62% classification accuracy on the CK+ corpus and a 77.84% classification accuracy on the UNBC corpus. This indicated a reduction of 18% and 17% accuracy from their approach.

Yuanye H. et al. [11] combined a bidirectional Long Short-Term Memory (LSTM) and a Convolutional Neural Network (CNN) model to determine the Emotion Intensity of tweets from an EmoInt-2017 task. The objective of the task was to determine the intensity of expressing an emotion in a tweet on a continuous numerical value. The LSTM architecture consisted of three gates plus a cell memory, which created a gating mechanism used to track sequence states. The CNN model consisted of two layers; convolutional and max pooling. For the dataset, word embeddings from GloVe, a pretrained word vector for Twitter containing 200 dimensions were used. The overall results obtained revealed that CNN alone had a 65.8% accuracy, LSTM had 56.1% while the combination of the two (BiLSTMCNN) had an overall accuracy of 67.7%.

Amol S. and Gerald M [12]. developed an affect recognition system that they used to estimate the affect intensity from multiple modalities. The multiple modalities comprised data captured from facial expressions, hand movements, body posture, and speech. Five participants enacted anger where the features of the different modalities were recorded in a text file. A scale of 0 and 1 annotation was used to label the intensity of the emotions, the aim being to estimate the classification rates of levels of emotion intensity from facial expressions, hand movements, body posture, and speech. For emotion prediction, an SVM classifier was applied, then the outcome was merged at the decision level before being allocated a level of confidence depending on the majority voting. Results for anger emotion detection from the multiple modalities were tabulated, followed by the results of intensity estimation. The combined results for all the modalities placed the model performance at an overall recall of 68%. Hand movements recorded the highest recall among the other modalities posting a recall of 61%, while speech recorded the least recall of 41%. However, on intensity estimation, the model recorded the highest accuracy performance from speech (73%) and the lowest accuracy performance on body movements (19%). On multiple modalities, the model achieved an accuracy performance of 46%.

3 Dataset

In the experiment, the Ryerson Audio-Visual Database of Emotional Speech and Song (RAVDESS) dataset was utilized [13]. This is a gender-balanced public dataset that is validated and obtained online at RAVDESS DATASET (Kaggle.com). Created by Ryerson University, Canada for the study of emotion, it is a database with 7356 files of emotional speech and song obtained from 24 recordings of 12 male and 12 female actors. One hundred four separate vocalizations of 60 spoken and 44 sung utterances were produced by the actors (A sample is shown in Table 1). Each vocalization was exported to give three separate modality states of Audio-Video, Video only, and Audio only. It yielded a total of 312 files from each actor, summing up to 7488 files. Because of some technicalities, 132 files from a female song actor were lost, leaving 7356 files to be used (4320 speech and 3036 song files). The actors uttered two distinct statements from the speech and song states twice intentionally using eight emotional conditions of disgust, fearful, happy, angry, sad, surprise, calm, and neutral.

Table 1 RAVDESS dataset sample

Row no.	Actor	Emotion state	Intensity	Actor number and path
0	Male	Neutral	Normal	1 Ravdess/Actor_1/03-01-01-01-01-01-01.Wav
1	Male	Neutral	Normal	1 Ravdess/Actor_1/03-01-01-01-01-02-01.Wav
2	Male	Neutral	Normal	1 Ravdess/Actor_1/03-01-01-01-01-01-01.Wav
3	Male	Neutral	Normal	1 Ravdess/Actor_1/03-01-01-01-02-02-01.Wav
4	Male	Calm	Normal	1 Ravdess/Actor_1/03-01-02-01-01-01-01.Wav
...	...	...	...	...
1435	Female	Surprise	Normal	24 Ravdess/Actor_24/03-01-08-01-02-02-24.Wav
1436	Female	Surprise	Strong	24 Ravdess/Actor_24/03-01-08-02-01-01-24.Wav
1437	Female	Surprise	Strong	24 Ravdess/Actor_24/03-01-08-02-01-02-24.Wav
1438	Female	Surprise	Strong	24 Ravdess/Actor_24/03-01-08-02-02-01-24.Wav
1439	Female	Surprise	Strong	24 Ravdess/Actor_24/03-01-08-02-02-02-24.Wav

1440 rows × 5 columns

The song utterances were deliberately voiced in six states of emotion; angry, calm, fear, happy, neutral, and sad. Five of these states were voiced in two levels of emotional intensity; strong and normal, leaving out neutral emotion state. The length of the utterances is seven syllables matched in word frequency and familiarity by use of MRC psycholinguistic database. The sung utterances were related to a piano sounded MIDI tones with fixed acoustic intensity, comprising six eight-note at 300 ms, ending with a quarter-note at 600 ms. Rating was done using three melody modes; major, neutral, and minor modes. The tonality of each melody relating to each emotion was made to be consistent with the emotional association. Two hundred forty-seven untrained research participants were used for rating, while a further 72 participants were used for test-retest data. A 9-point valence scale was used to rate a perceived valence of 3 melody modes from a self-assessment manikin (SAM). Three main epochs; Task presentation, counting, and vocalization marked the stimulus timeline. Task presentation epoch was done at 4500 ms, counting epoch at 2400 ms, and vocalization epoch at 4800 ms, giving a total of 13,700 ms stimulus trial time.

4 Methodology

The methodology followed in the experimental setup was first to extract features from the audio dataset, separate Acoustic features from the Lexical features, and then perform feature selection [14]. The Librosa Python library was used to extract the audio features from the dataset. The extracted features from the audio dataset are Mel Frequency Cepstral Coefficient (MFCC), Mel, and Zero Crossing Rate (ZCR). Features derived from the selection were intended to inform and facilitate the learning and generalization of the model for a better understanding of the human interpretation. Features were then classified according to their domains. The Time domain features such as Mel Frequency Cepstral Coefficients (Table 2) were obtained from MFCC and the Frequency domain features were obtained from pitch. Next, we labeled the data, assigning each variable as shown in Table 3.

The selected features were split into training and testing sets, with training dataset taking 80% of the whole dataset while testing took 20% of the dataset. These features were then introduced to a model and a trained classifier as shown in Fig. 1.

Table 2 A Mel spectrogram

Row no.	Mel frequency cepstral coefficient feature
0	[−76.38477, −76.38477, −76.38477, −76.38477, − . . . .
1	[−75.33552, −75.44532, −75.55403, −75.20395, − . . . .
2	[−75.15071, −75.15071, −75.15071, −75.15071, − . . . .
3	[−75.26845, −75.26845, −75.26845, −75.26845, − . . . .
4	[−80.14738, −80.14738, −80.14738, −80.14738, − . . . .

Table 3 RAVDESS coding

Identifier	Coding description of factor levels
Emotion	01 denotes "neutral," 02 denotes "calm," 03 denotes "happy," 04 denotes "sad," 05 denotes "angry," 06 denotes "fearful," 07 denotes "disgust," and 08 denotes "surprised"
Intensity	01 denotes "Normal" and 02 denotes "strong"
Statement	The phrases; "kids are talking by the door" are represented by 01 and "dogs are sitting by the door" is represented by 02
Repetition	01 represents "first repetition," 02 represents "second repetition"
Actor	First actor . . . 24 twenty-fourth actor indicated by 01

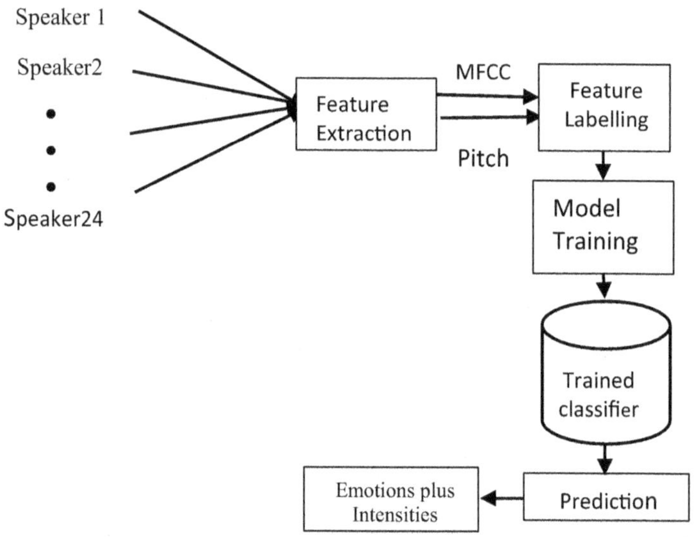

Fig. 1 System design pipeline

5 Experiments

The process begins by extracting the appropriate data features using MFCC feature extraction algorithm (Fig. 2). First, the **analog audio data was converted to a digital signal** to enable acoustic features to be extracted. The compressed energy from the conversion is produced at the **preemphasis** at a high energy level. This is done so as to balance the voiced sound spectrum with a high roll-off in the high-frequency region. At the **windowing** stage, tracking of individual speech temporal characteristics of sound is done to obtain short-term spectral measurements. It is normally passed through a 20-ms window at an interval of every 10 ms window.

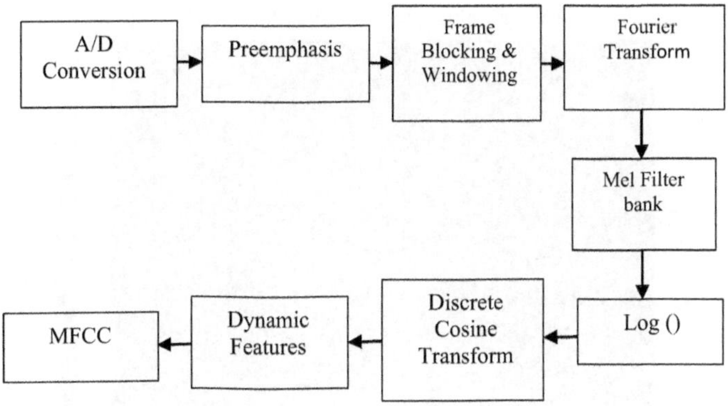

Fig. 2 MFCC feature extraction process

To change the sound signal from time domain to frequency domain, **Discrete Fourier Transform** is applied. This converts the signal into a magnitude spectrum by applying a mathematical function:

$$X(k) = \sum_{n=0}^{n\_1} x(n)e\frac{-j2\pi nk}{N}; o \leq k \leq N - 1$$

where; x(n) represents a long analysis window (hamming) of N samples and K represents the length of DFT.

Mel spectrum is calculated by passing the Fourier transformed signal through a set of bandpass filters called Mel filter bank. The short-term energy of the actual logarithm is displayed as a discrete cosine transform on the Mel frequency scale. With the help of a filter bank, MFCC is grounded on a signal disintegration which can be converted from a real frequency (f) to a Mel frequency Mel(*f*) by using the following equation:

$$Mel(f) = 2595\ x\ \log_{10}(1 + f/700)\,.$$

where:

Mel(f) = Perceived frequency in Mels.

f = Physical frequency in Hertz (Hz).

The **logs of feature** energies from the Mel filter banks are obtained. These energies are a result of a compression operation that scales down the features to closely mimic the human actual hearing mechanism. The logarithms also allow the use of a channel normalization technique known as cepstral mean subtraction. Finally, the computation of the **DCT** log filter banks energies is done. This removes the correlation of the filter banks due to their overlapping nature. The DCT produces 26 coefficients which only 12 are used in the experiment. The higher DCT coefficients are removed to prevent the degradation of automatic speech recognition

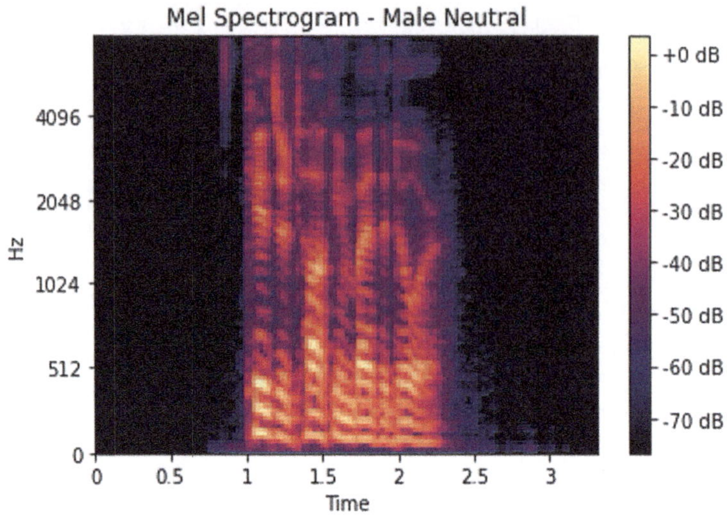

Fig. 3 Mel spectrogram visualizes the sinusoidal wave of a male voice with neutral emotions as an MFCC feature

performance due to their fast-dynamic nature. The short-term representation of sound power spectrum known as Mel Frequency Cepstrum (MFC) was selected as the ideal feature for this experiment. The frequency bands in MFC are spaced equally when placed on a Mel scale. The Mel scale closely approximates the response in a human auditory system. Specifically, the Mel Frequency Cepstral Coefficient (MFCC) were selected due to its wide use in audio classification experiments and good performance [15]. The principle used to obtain the feature characteristics was the Mel Cepstral which obtains the short-time spectral shape containing valuable data describing the quality of voice and production effects (Fig. 3). The coefficients are obtained by calculating the cosine transform of the real logarithm of the short-term spectrum of energy.

The selected features were split into training and testing data before being introduced into the model. The model is a one-dimension (1D) CNN with 8 layers (Convolutional1&2, Max-Pooling1&2, Dropout1&2, Flatten and Dense). A sigmoid activation function was introduced in the dense layer for the purpose of classifying the emotions and categorical cross entropy was used to optimize and determine the model accuracy and loss. A training batch size of 32 was used to train and validate the model after 40 epochs. Experiments were done on the dataset in five scenarios. Scenario I was the actual detection of variables for eight emotion categories only (Table 4). Scenario II involved combining the eight emotions with their categories (Table 5). Scenario III involved omitting two emotions (Calm and Neutral) and combining the remaining six emotions and their intensities as variables (Table 6). In Scenario IV we used four emotions (Angry, Fear, Happy, and Surprise) and their intensities as variables (Table 7). In Scenario V we used the remaining three

Table 4 Emotions outcome

	Precision	Recall	F1-Score	Support
Angry	0.52	0.62	0.56	39
Calm	0.54	0.82	0.65	38
Disgust	0.64	0.73	0.68	37
Fear	0.69	0.24	0.35	38
Happy	0.38	0.31	0.34	35
Neutral	0.34	0.50	0.34	24
Sad	0.30	0.16	0.21	38
Surprise	0.50	0.59	0.54	39
Accuracy			0.50	288
Macro average	0.49	0.49	0.47	288
Weighted average	0.50	0.50	0.47	288

Table 5 Combined emotion and intensity outcomes

	Precision	Recall	F1-Score	Support
Angry/Normal	0.40	0.40	0.40	20
Angry/strong	0.53	0.89	0.67	19
Calm/Normal	0.33	0.16	0.22	25
Calm/strong	0.35	0.85	0.50	13
Disgust/Normal	0.38	0.52	0.44	21
Disgust/strong	0.27	0.38	0.32	16
Fear/Normal	0.57	0.22	0.32	18
Fear/strong	0.59	0.65	0.62	20
Happy/Normal	0.23	0.31	0.26	16
Happy/strong	0.33	0.21	0.26	19
Neutral/Normal	0.37	0.46	0.41	24
Sad/Normal	0.38	0.19	0.25	16
Sad/strong	0.60	0.14	0.22	22
Surprise/Normal	0.23	0.16	0.19	19
Surprise/strong	0.22	0.25	0.23	20
Accuracy			0.38	288
Macro average	0.39	0.39	0.35	288
Weighted average	0.39	0.38	0.35	288

emotions (Calm, Disgust, and Sad) with their intensities as variables (Table 8). In all the five scenarios, a procedure was adopted where individual variables were introduced to the model separately. The detection of emotions and intensities were done separately and thereafter categorization of the emotions and intensities was obtained from the classification model.

Table 6 Combined emotions and intensity outcome after removing calm and neutral emotions

	Precision	Recall	F1-Score	Support
Angry/Normal	0.41	0.45	0.43	20
Angry/strong	0.55	0.57	0.56	21
Disgust/Normal	0.39	0.39	0.39	23
Disgust/strong	0.38	0.33	0.36	15
Fear/Normal	0.33	0.26	0.29	19
Fear/strong	0.33	0.24	0.28	17
Happy/Normal	0.31	0.29	0.30	17
Happy/strong	0.42	0.21	0.28	24
Sad/Normal	0.26	0.69	0.38	16
Sad/strong	0.25	0.05	0.08	21
Surprise/Normal	0.35	0.61	0.45	18
Surprise/strong	0.26	0.25	0.26	20
Accuracy			0.35	231
Macro average	0.35	0.36	0.34	231
Weighted average	0.36	0.35	0.34	231

Table 7 Outcome of four emotions and their intensities

	Precision	Recall	F1-Score	Support
Angry/Normal	0.42	0.54	0.47	24
Angry/strong	0.64	0.93	0.76	15
Fear/Normal	0.40	0.09	0.14	23
Fear/strong	0.54	0.39	0.45	18
Happy/Normal	0.36	0.40	0.38	20
Happy/strong	0.35	0.35	0.35	17
Surprise/Normal	0.40	0.75	0.52	16
Surprise/strong	0.29	0.19	0.23	21
Accuracy			0.43	154
Macro average	0.42	0.46	0.41	154
Weighted average	0.42	0.43	0.40	154

Table 8 Outcome of three emotions and their intensities

	Precision	Recall	F1-Score	Support
Calm/Normal	0.36	0.42	0.39	19
Calm/strong	0.65	0.74	0.69	23
Disgust/Normal	0.62	0.38	0.47	21
Disgust/strong	0.50	0.81	0.62	16
Sad/Normal	0.56	0.43	0.49	21
Sad/strong	0.77	0.62	0.69	16
Accuracy			0.56	116
Macro average	0.58	0.57	0.56	116
Weighted average	0.58	0.56	0.56	116

6 Results

6.1 Scenario I: Emotion Detection from the Eight Emotion States

After training and testing our model on the data, using the eight emotional states (angry, calm, disgust, fear, happy, neutral, sad, and surprise) as variables, it indicated an overall performance accuracy of 50% (Table 4). The model indicated a good performance in detecting calm, disgust, angry, and surprise; posting a recall of 82%, 73%, 62%, and 59%, respectively. Others posted a recall of 50% and below, with sad emotion having a recall of 16%.

6.2 Scenario II: Emotion and Intensity Detection Using both Emotions and Intensity States as Variables

In this stage, the emotions and intensity variables were combined (Table 5). The model's overall performance dropped to 38% from the previous 50%. Of the eight emotions combined with their intensities, the model posted a good recall on only four variables (Angry Strong at 89%, Calm Strong at 85%, Fear Strong at 65%, and Disgust Normal at 52%). The rest of the variables posted a recall of less than 50%, with Sad Strong posting the least outcome of 14%.

6.3 Scenario III: Emotion and Intensity Detection After Omitting Calm and Neutral Variables

During this stage, the variables calm and neutral were omitted due to the shared similarity of characteristics and neutral having only one intensity variable; normal (Table 6). The model performance dropped after omitting the two variables (calm and neutral). Out of 12 variables, only three variables posted a recall of over 50%. The model had a recall of 69% on Sad Normal, 61% on Surprise Normal, and 57% on Angry Strong. The overall model performance dropped again to 35% from 38%.

6.4 Scenario IV: Emotion and Intensity Detection Using Angry, Fear, Happy, and Surprise as Variables

After omitting other variables and remaining with only four, our model exhibited an improvement in performance. It posted an overall accuracy performance of 43%

(Table 7), an improvement of 8% from the previous performance. The best recall posted was 93% for the variable Angry Strong and the least recall attained was 09% for the variable Fear Normal.

6.5 *Scenario V: Emotion and Intensity Detection Using Calm, Disgust, and Sad as Variables*

The last stage was to introduce the three remaining variables to the model and determine the outcome (Table 8). The highest recall attained by the model was 81% on the variable Disgust Strong and the least recall attained was 38% on the variable Disgust Normal. The model's overall performance improved and attained a recall accuracy of 56%, an improvement of 23% recall accuracy.

7 Discussions

The CNN model we proposed achieved an overall best performance of 56% recall as shown in Table 9. This was after performing a classification on three opposing emotions using a single classifier (Logistic Regression). On emotion classification alone, the model attained a recall accuracy of 50% on an output of eight emotion classes. The model performance dropped to 38% when classification was done on combined emotion and intensity variables. In scenario III, we omitted two emotions with similar characteristics (Calm and Neutral) and performed a classification on the remaining six basic emotions [3]. The result was a drop in the model performance achieving a recall of 35%. In scenario IV, we omitted four classes of emotions (Calm, Neutral, Disgust, and Sad) and did a classification on four classes; two opposing and two with similar characteristics. The model attained an improvement of 8% to post a recall of 43%.

Table 9 Model performance

Scenario	Variables	Model overall performance (%)
I	Emotion detection from the eight emotion states	50
II	Emotion and intensity detection using both emotions and intensity states	38
III	Emotion and intensity detection after omitting calm and neutral	35
IV	Emotion and intensity detection using angry, fear, happy and surprise	43
V	Emotion and intensity detection using calm, disgust and sad	56

Based on the experiment results, it is evident that determining the intensities of emotions from audio dataset requires a multi-modal approach in order to achieve the best results [8]. This is informed by the many similar characteristics exhibited by each individual emotion. Emotions can be discreet or continuous [4]. They are also determined by the frequency spectrum range where they lie. Determining intensities of emotions from opposing frequency spectrum ranges using appropriate approaches can yield good results. This is evident in our experiment's scenarios IV and V where in scenario IV, four opposing emotions (two with similar characteristics) were grouped together and their intensities determined. The output was a drastic improvement from the previous outcome of 35% accuracy to 43% accuracy. In scenario V, the intensities of three opposing emotions were determined and the outcome was an improvement of 13% from scenario IV the model posting an overall accuracy of 56%.

8 Conclusion and Future Work

In our study, we used a CNN to determine emotion intensities on the RAVDESS dataset. A Logistic Regression classifier was used to classify MFCC features to be introduced to the model. Emotions were first classified separately and then their intensities were determined. The model achieved an accuracy of 50% when emotions alone were classified. It achieved the lowest accuracy of 35% when both emotions and intensities were determined at the same time. When the intensities of three opposing emotions were determined, the model achieved a high accuracy performance of 56%.

We applied a single classifier on the dataset in our experiment, to classify individual emotions and their intensities separately. Parallel and serial models with two or more classifiers [8] were other available options for determining emotions. This will be explored in our future work on this study.

References

1. How sound is produced by Humans. (2019). Retrieved July 2, 2022, from https://byjus.com/physics/sound-produced-humans/
2. Caschera, M. C., Grifoni, P., & Ferri, F.. *Emotion classification from speech and text in videos using a multimodal approach*. Retrieved July 2, 2022, from https://www.mdpi.com/journal/mti
3. Abhijit Mondal: *Mining emotions on Plutchik's Wheel*. (2020). Conference paper 2020 Seventh international conference on social networks analysis, Management and Security (SNAMS)
4. Bota, P. J., Wang, C., Ana, L. N. F., & da Silva, H. P. (2019). A review, current challenges, and future possibilities on emotion recognition using machine learning and physiological signals. *IEEE Open Access Journal, 7*, 140990–141020.
5. Paul, W. (2021). *Flint: Laryngeal and pharyngeal functions* (2nd ed.).

6. Poria, S., Majumder, N., Mihalcea, R., & Hovy, E. *Emotion recognition in conversation: Research challenges, datasets, and recent advances*. Michigan Institute for Data Science, US. Retrieved July 4, 2022, from https://www.researchgate.net/publication/332960306_Emotion_Recognition_in_Conversation_Research_Challenges_Datasets_and_Recent_Advances
7. Avetisyan, H., Bruna, O., & Holub, J. (2016). Overview of existing algorithms for emotion classification. Uncertainties in evaluations of accuracies. *Czech Technical University in Prague. November 2016 Journal of Physics Conference Series, 772*(1), 012039.
8. Daube, C., Ince, R. A., & Gross, J. (2019). Simple acoustic features can explain phoneme-based predictions of cortical responses to speech. *Current Biology, 29*(12), 1924–1937.e9. Published by Elsevier Ltd (2019).
9. Kawase, M., & Nakayama, M. (2021, October 6–8). *Emotion intensity level analysis of speech emotional intensity estimation*. ESANN 2021 proceedings, European Symposium on Artificial Neural Networks, Computational Intelligence and Machine Learning. Online event, i6doc.com publ., ISBN 978287587082-7.
10. Thuseethan, S., Rajasegarar, S., & Yearwood, J. (2019). Emotion intensity estimation from video frames using deep hybrid convolutional neural networks.
11. He, Y., Yu, L. C., Lai, K. R., & Liu, W. (2017). *Determining emotion intensity using a bi-directional LSTM-CNN model*.
12. Amol, S., & Gerald, M. (2016). *Affect intensity estimation using multiple modalities*. Proceedings of the 27th International Florida Artificial Intelligence Research Society Conference, FLAIRS 2014 ER.
13. Livingstone, S. R., & Russo, F. A. (2018). Retrieved July 2, 2022, from). *The Ryerson audio-visual database of emotional speech and Song (RAVDESS)*: A dynamic, multimodal set of facial and vocal expressions in north American English. *PLoS One, 13*(5), e0196391. https://doi.org/10.1371/journal.pone.0196391
14. Kim, E., Song, H., & Shin, J. W. (2020). Affective latent representation of acoustic and lexical features for emotion recognition. *Sensors (Basel), 20*(9), 2614. Published online 2020 May 4. https://doi.org/10.3390/s20092614
15. Alang Rashid, N. K., Alim, S. A., & Hashim, S. W. (2017). Receiver operating characteristics measure for the recognition of stuttering dysfluencies using line spectral frequencies. *IIUM Engineering Journal, 18*(1), 193–200.

Machine Learning-Based Forward Collision Avoidance System: A Case Study for the Kayoola EVS

Ali Ziryawulawo, Adonia Mbarebaki, and Sam Anael

1 Introduction

Driving safety is one of the key considerations in the development of autonomous vehicles. Autonomous vehicles are the promising next-generation transport means, they are able to see and understand the road and the obstacles on it in order to be able to maneuver on the road and achieve the approximate human driver driving or even drive better. All this is due to the nascent artificial intelligence and Machine Learning [1]. To realize these benefits promised by autonomous vehicles, reliable and efficient machine learning algorithms are required. This can be effectively achieved using the reinforcement learning type of machine learning for autonomous cars and other ADAS like Open pilot for autonomous cars. These systems can achieve up to level 4 standard of autonomous vehicles according to SAE [2].

The industrial strategy for automotive safety systems has been evolving for the last 20 years. Initially, individual passive devices and features such as seatbelts, airbags, knee bolsters, and crush zones [3] were developed to save lives and minimize injuries when an accident occurs. The automotive industry is now at the stage of actively avoiding accidents as well as providing maximum protection to the vehicle occupants and even pedestrians. The systems that are under intense development include collision avoidance systems [4].

A. Ziryawulawo (✉) · S. Anael
The Nelson Mandela African Institution of Science and Technology, Arusha, Tanzania
e-mail: ziryawulawoa@nm-aist.ac.tz

A. Mbarebaki
Kiira Motors Corporation, Kampala, Uganda

J. Marx Gómez et al. (eds.), *Artificial Intelligence Tools and Applications in Embedded and Mobile Systems*, Progress in IS,
https://doi.org/10.1007/978-3-031-56576-2_13

Kiira Motors Corporation is built on a premise to champion Value Addition in the domestic automotive industry for job creation and diversification of Uganda's economy. The Kayoola EVS buses represent the next generation of public transport for cities within Uganda and the East African region at large [5]. The proposed Forward Collision Avoidance system will engage brakes when it meets an obstacle in front of it for safe motion without the driver's intervention or with little intervention. As this will not automatically dislodge the driver, the driver is supposed to keep the eyes on the road and provide manual controls as a backup to the system. The driver takes action for safe motion by either braking or steering away. This system uses the image array and collision events as inputs to the machine learning algorithm in order to be able to maneuver along the highway without colliding with other vehicles [6].

Autonomous vehicles have the potential to provide a wide range of traffic solutions. Automotive technology is undergoing a shift where active safety has become more of concern [7]. Autonomous vehicles are able to see and understand the road, maneuver around obstacles, and be able to achieve approximate human driving and even better with time [8]. With the nascent artificial intelligence and machine learning, algorithms for autonomous vehicle systems have been developed and deployed, especially the forward collision avoidance system [9–12].

For this project, a machine learning collision avoidance system is designed based on reinforcement learning [13] in order to determine the preset obstacles by interacting with the environment and be able to achieve a reliable and efficient algorithm. A study of the rewards and penalties as regards to collisions is carried out together with CARLA simulator simulations on the performance of the algorithm. An optimized deep Q-learning algorithm that relies on deep reinforcement learning was developed under constrained conditions in a Carla simulation environment with a metric loss of 20%. The algorithm relies on dynamic programming which can buffer data during the training process. A unique stratified collision avoidance system is developed to manage an optimal self-driving environment. The system delivers an effective self-driving framework that meets the intended goals by utilizing deep reinforcement learning. This proposed system offers an efficient deep reinforcement learning system for learning driving behaviors that share the acquired driving patterns with neighbors through parameter distribution to maximize resource use and exchange learning experiences across agents.

2 Related Works

The authors in [14] suggest a straightforward solution based on monocular vision to solve the data scarcity issue and advance forward collision avoidance systems. To predict the velocity of the ego-car and to detect the leading car, they use cutting-edge neural network-based optical flow and monocular depth prediction algorithms in addition to object identification.

In order to avoid object recognition and depth sensing, Kilicarslan et al. [15] estimated time-to-collision using motion divergence to identify incoming objects.

The method was evaluated in diverse situations. The movement profiling technique in this study acquires object motion, disregard the majority of background objects, and is more reliable compared to optical flow, except that it is susceptible to errors because of false motion caused by shadows and reflections, camera trembling on non-uniform roads, and camera tilting in scenarios involving car braking. Wulfe et al. [16] examined real-time collision estimation based on simulations to assist minimize challenges caused by the scarcity of collisions as well as the requirement for vast volumes of data. The study defines risk prediction as policy analysis inside the framework of the Markov decision process.

Venkateswaran et al. [17] propose a unique forward collision warning model that uses monocular vision with range projection, in which moving cars are recognized initially with a pre-trained YOLO algorithm. These cars are then issued an identifier and traced with the Hungarian algorithm and the Kalman filter. The distance from the identified car to the ego vehicle is then determined by inverse perspective mapping of the image into a bidimensional orthogonal top-down view.

Authors in [18] propose a method based on the combination of camera and LIDAR sensors. The camera was utilized for target identification and localization, whereas the LIDAR sensor provided 3D data for vehicle proximity prediction. A spatial measurement was employed to translate the 3D LIDAR data to the 2D image domain. This approach was designed as a demonstration leveraging the open-source Autoware platform. Shi et al. [19] developed a hardware friendly time-to-collision prediction approach based on dense optical flow field divergence. They computed optical flow using motion energy characteristics using random forests.

3 Methodology

3.1 Sequential Model Design

The system model design was subdivided into the design of agent code on how to interact within the environment and then also an environment modelled with specific features like other vehicles and cyclists.

Environmental Setup

The project was executed in the Anaconda virtual environment with different libraries required for running the simulator and the machine learning system were installed, which includes keras with tensor flow backend, NumPy, and matplotlib for calculus, pygame for visualization during the training process of the algorithm.

Data Processing

The simulator has 200 available different vehicle types which include trucks, cars, shuttles, and motorbikes. Given the system uses a bus, an approximate of a bus which is a shuttle was modelled and the sensors attached to the dashboard.

The system was initialized with replay memory that stores frames of images during the driving process and from that memory the system samples with a batch size of 64 per second during the running of the simulation. The camera preview during the training process was set to True to have a visual view. The RGB camera was used where the image is preprocessed to three channels. The collision sensor was attached to record collision events during the training process.

Transitions of video frames are put into a buffer and sample a mini-batch of samples of size 16 from this buffer to train the deep network. This forms an input dataset that is stable enough for training. As we randomly sample from the replay buffer, the data is more independent of each other [20, 21].

Two deep networks that is model and target model. The first one is used to retrieve the Q-values while the second one includes all updates in the training. After running the updates in the Q-table, model with target model is synchronized [21]. The purpose is to x the Q-value targets temporarily, so we do not have a moving target to chase. In addition, parameter changes do not impact target model immediately [22]. With both experiences replay and the target network, a more stable input and output to train the network, and behaves more like supervised training.

Target Network Initialization

Two deep networks, that is, model and target_model are created. The first one is used to retrieve the Q-values while the second one includes all updates in the training. After running the updates in the Q-table, model with target_model is synchronized [21]. The purpose is to x the Q-value targets temporarily so we do not have a moving target to chase. In addition, parameter changes do not impact target_model immediately [22]. With both experiences replay and the target network, a more stable input and output to train the network, and behaves more like supervised training.

Designing the Convolutional Neural Network for the System

The first convolutional layer was designed with 32 input nodes and it receives a sequence of images with three channels, that is, red, green, and blue images of size (640,480). Rectified linear unit function was used as an activation function and the maxpooling layer was set to have a pool size of (2,2) with a percentage drop out of 20% in the data received without losing the information required. The second consecutive convolutional layer was designed to receive image arrays processed by the first convolutional layer and the processing was done and also 20% dropout was initialized. The Flattening layer was added to receive outputs from the convolutional neural network inform of 3D features and preprocesses it to 1D feature vector. The 1D output of the feature vector was fed into the dense layer with 64 input nodes from which a prediction is made. The outputs of the 64 input nodes are fed into another dense layer with three output nodes from which the driving commands are predicted from. The reinforcement learning model design is shown in Fig. 1.

3.2 Rendering with Carla Simulator

CARLA is an open-source photorealistic simulator [20] developed to train, validate, and test autonomous driving algorithms. It provides digital assets and complete control of actors on the map, environmental conditions control, a sensor suite, maps generation, a flexible API, and a server-client-based communication. In addition to CARLA Simulator, which embeds all the control logic, the rendering, the physics and all the actor properties, CARLA offers a Python API module. The server is the Simulator and the client-server communication is controlled through the Python API as shown in Fig. 2.

Most of the aspects of the simulation are accessible from the Python API. With Python scripts, it is possible to retrieve raw data coming from CARLA sensors attached to the ego vehicle, process them, compute all the parameters needed by

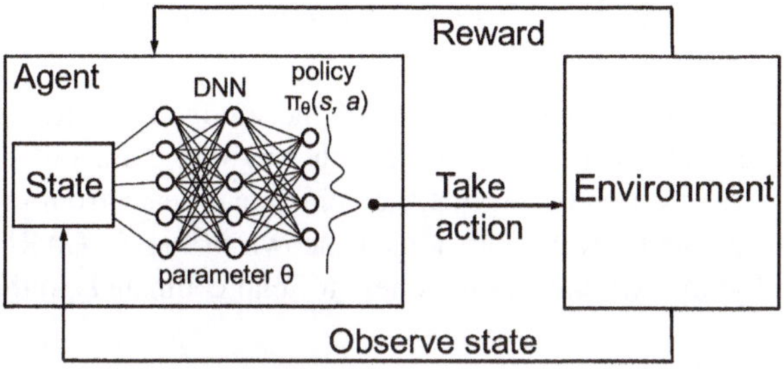

Fig. 1 Reinforcement learning model design

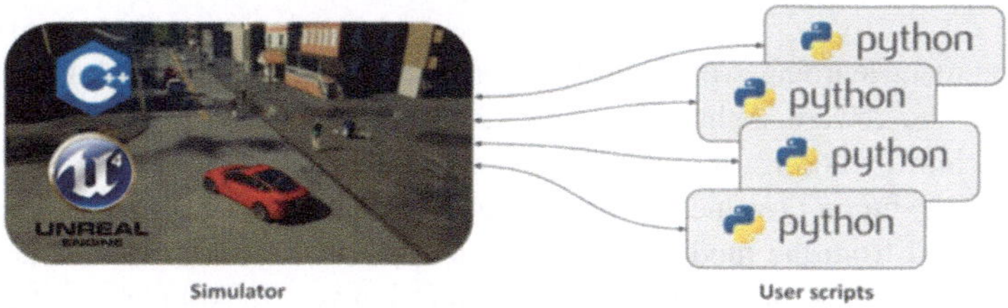

Fig. 2 CARLA server-client communication

the controller, and send to CARLA Simulator the controls of throttle, brake, and steering. Each Python script from the client side can be logically divided into two different parts:

Configuration of the Simulation and of the Actors

Before starting the control algorithm, the connection with the server and all the settings of the simulation and the actors have to be done. Here, information on the world can be retrieved, actors can be spawned at arbitrary locations and orientations, sensors can be attached to the ego vehicle and other actors can be created, spawned, or destroyed.

Client-Server Synchronized Communication

The control algorithm is contained in this part, where the proper simulation is started. The sensor(s) attached to the vehicle produce raw data and the client subscribes to the sensor stream by providing a callback function that is called each time a new data is generated by the sensor. This callback function is the one that stores the images in a queue accessed by the control algorithm to return the controls to the vehicle. At the end of the simulation the client disconnects [23].

3.3 Reinforcement Learning with Carla Simulator

In Forward Collision Avoidance System of the autonomous Kayoola EVS, the reinforcement learning will be used in a such way that collision is not rewarded and no collision is rewarded such that the vehicle can learn from the policy for a long time and be able to avoid collisions as shown in Fig. 3. CARLA acts as the server and the Python API used as the client to send commands to the actor within the simulator.

Action-State Rewards Based on Vehicle Speed and Distance

Reinforcement learning works on the principle of action-state reward strategy to maximize the reward. When the car (Agent) is in an environment, it interacts with it while exploring the environment using the epsilon greedy policy which picks the maximum Q-value. Q-value stands for quality value. These Q-values are approximated from the model and from which predictions are made. The epsilon greedy policy works on the principle of exploration versus exploitation, it explores the environment and after picking the maximum Q-value it uses the maximum Q-value to predict future Q-values [24]. This process leads the agent learning to

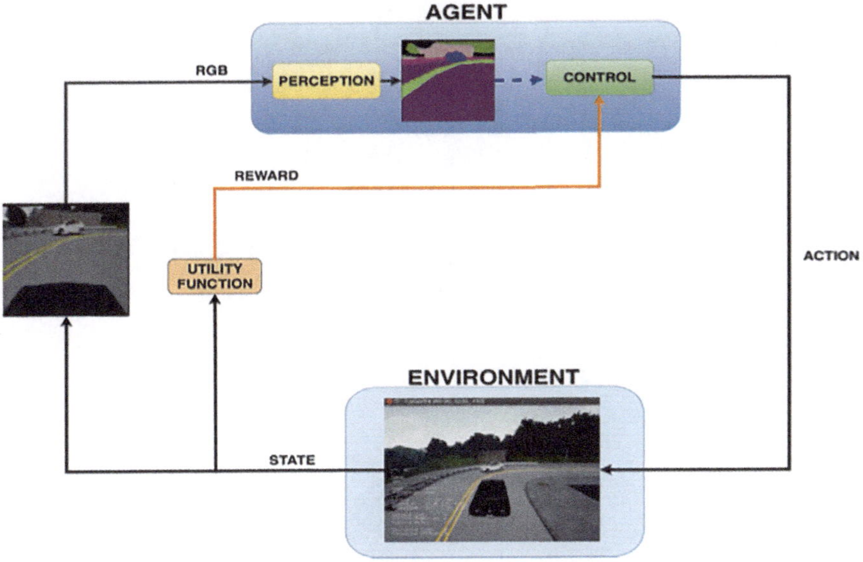

Fig. 3 Reinforcement learning in CARLA

interact with the environment and training takes for different episodes and the learnt strategy is visualized in a simulation demonstration or even more quantitative graphs on the Tensor board showing the maximum rewards.

Forward Collision Avoidance System Flowchart

Figure 4 shows the CARLA collision avoidance flowchart. During the time of simulation of a car moving the threshold speeds are initialized and also the threshold distance.

Two models are created where one is used to take in all raw values while the other model which acts as the target model network is the one used for actual predicting of the learnt strategy. For this reinforcement learning strategy, the rewards are based on the collision of the agent with obstacles in the environment, speed of the agent, and the distance from other the immediate vehicle in front of it. When the vehicle is driving within the environment and it collides with an obstacle, the system receives a penalty of -1 to demotivate the agent from the collision and 1 for any action that does not lead to a collision like driving in a straight line and maintaining a threshold distance of 2 m. A 60-km/h forward speed is initialized to actuate the car in a forward motion. When the car controls and speed are not initialized normally, the vehicles do not discover which action is the best to take and drive in circles. 60 km/h speed was initialized in a way that when the speed is greater than the set threshold, it terminates the episode of training and when it is less than 60 km/h it receives a reward of 0 and when it is less than 60 km/h it receives a reward of 0 [25].

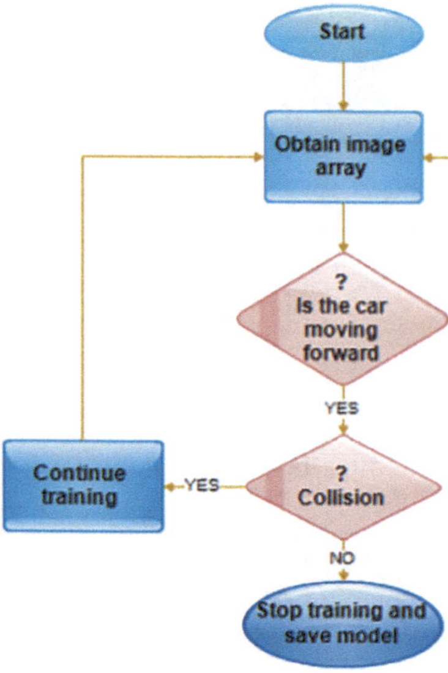

Fig. 4 CARLA collision avoidance flowchart

3.4 *Training of the Sequential Model*

The neural network used for the DQN consists of fully connected layers with five hidden layers. Adam Optimizer algorithm is used to minimize the loss with learning rate = 0.0001 [26].

The number of batch size data samples used as a state is set to n = 16. We set the size of the replay memory to 5000. The training was done initially for 1000 episodes, then increased to 1500 episodes for 16 h, and finally for 3000 episodes for approximately 40 h. Sequential model was created which is able to steer the vehicle in forward motion and brake for cases of imminent collision.

Deep Q-Network was trained by letting the vehicle agent drive within the simulation environment (CARLA Simulator). Throughout the training of the DQN, an action has to be selected for every given input image. For exploration purposes, an epsilon greedy policy is used with epsilon giving the probability for selecting an action randomly. Otherwise, the DQN estimates the Q-values for every action given the input image [25]. The action with the maximum Q-value is selected. At the start of the training epsilon is set to 1, but it is linearly reduced with an increasing number of iterations until a minimum value is reached after 1500 episodes of iteration.

Training was later increased to 3000 episodes It was observed that the vehicle collisions in the system reduced as the loss profile continued to reduce as observed from the loss profile graph.

4 Results Analysis

This chapter contains the key results obtained, the key assumptions and simulation learning parameters used to attain the results. The visualization graphs for model training.

4.1 Sequential Model Results

Training the Sequential Model

Initially as shown in Fig. 5, the training was done with 1000 episodes and these saturated at around the 300th episode, the model accuracy was very high at around 95%, and normally in reinforcement learning with deep q-networks this accuracy is due to overfitting. That is the process of taking in very many inputs and generalizing on a few of them. This leads to a higher accuracy but practically which is not possible.

Sequential Model Optimization

The training was again done with a dense layer added to reduce the overfitting problem and the results improved in performance. The performance of the agent in a simulated environment increased considerably compared to the training for the first

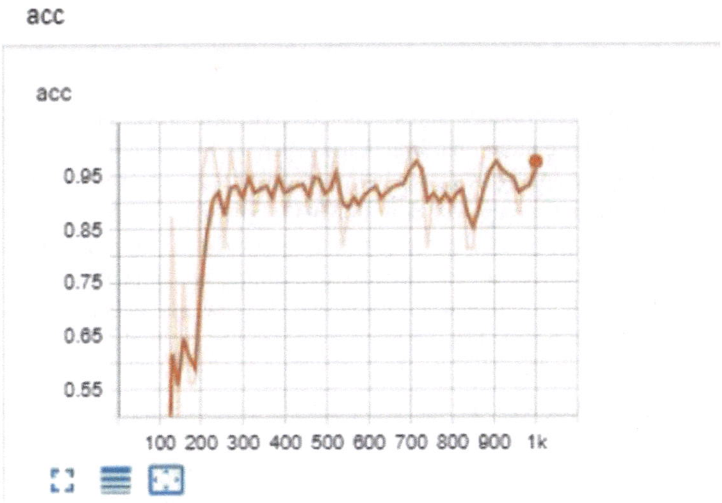

Fig. 5 First case sequential model training for 1000 episodes

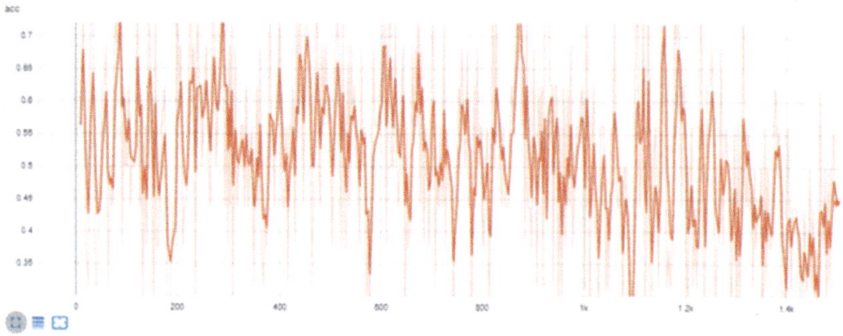

Fig. 6 Accuracy of model training for 1500 episodes

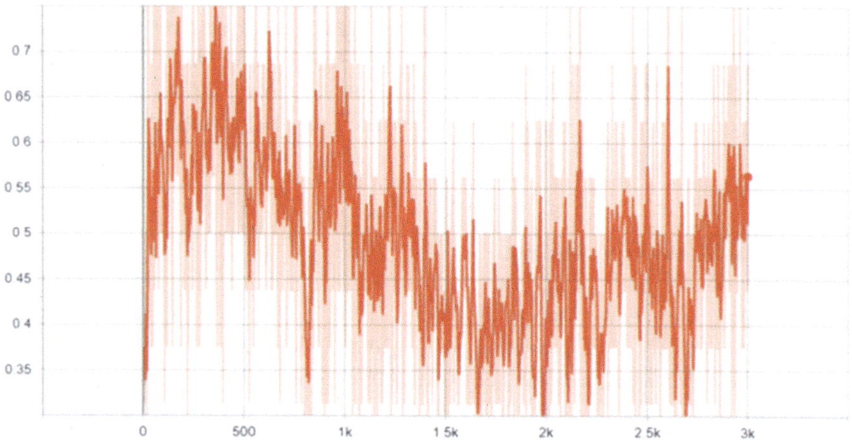

Fig. 7 Accuracy for 3000 episodes

1000 episodes the performance in a simulated environment was better compared to the first training before addition of the dense layer.

The percentage accuracy as shown in Figs. 6 and 7 was observed to increase from 1500 episodes when training was done for 3000 episodes which was translated to the simulator and the agent could execute the slow down action and stop in case of an obstacle in front of it. The process of training was repeated with different environmental conditions like when rainy and the rewards, accuracy, and loss of the model were computed. Fig. 6 shows an episodic graph showing the rewards against training episodes. For the deep Q reinforcement learning algorithm, the accuracy values are not very high since the agent takes in random actions and makes a prediction depending on the inputs. It was observed that the loss metrics are observed to reduce which implied the agent was learning from the environment. The accuracy metrics are observed to average at around 60%. This accuracy is acceptable for deep Q-learning networks.

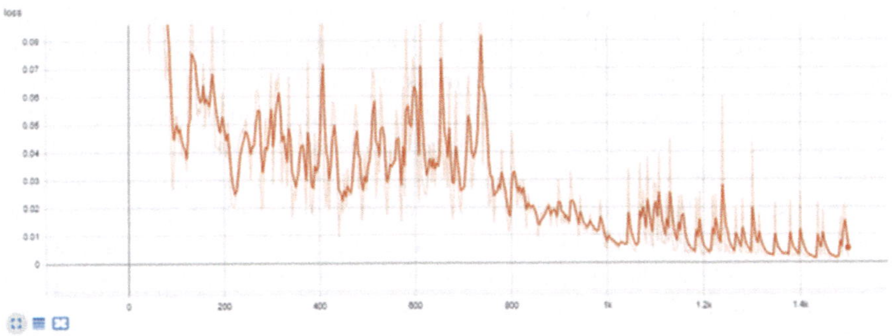

Fig. 8 Accuracy of model training for 1500 episodes

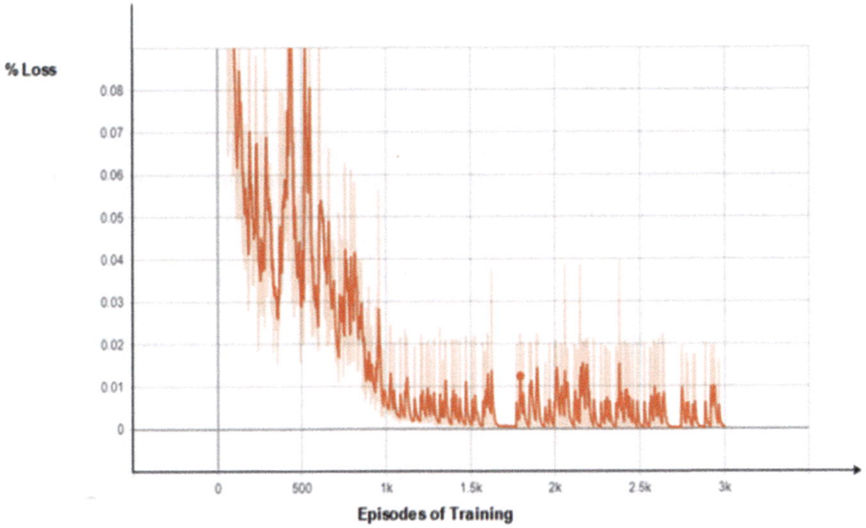

Fig. 9 Loss for 3000 episodes

Loss Profile for the Sequential Model

It was observed that during the training process, the loss profile was reduced as shown in Figs. 8 and 9 and this signifies that the agent was learning to act within the environment according to commands from the reinforcement learning model.

4.2 Xception Model Results

Xception model was pre-trained on ImageNet dataset with approximately 90 classes. It is from these different classes that the agent is able to differentiate between obstacles while exploring the environment. Sixty-four input nodes for predictions are used since the same driving commands are used in either model, the reward strategy in this model was given in a such way that a collision was penalized with a very huge negative reward, −500 to demotivate it and a no collision with a 1 to motivate the action taken within the environment. A value of −1 reward was given for a car not exceeding 50 km/h that is driving within the threshold speed.

The results of the training process were obtained. It was observed that over 100 episodes of training the agent learnt that whenever a collision would occur it would get back on track and the vehicle would accelerate very slowly and when a collision occurs it again gets back on track.

The training episodes were increased to 850 episodes and it was realized that the learning process was minimal though there was a slight improvement in the distance travelled without knocking other obstacles in the environment.

Results of Xception Model

It is observed from the graph in Fig. 10 that the average reward is about −2 and saturated after 600th episode onward which showed that there was no significant learning after those episodes. This model could not be relied on since it had a lot of inefficiencies like rampant collisions and getting off the road track.

Figure 10 shows the variation of average reward as the training goes on against episodes. The average reward was observed to be around −2. According to the initialization, this implies collision is imminent as the agent traverses the environment.

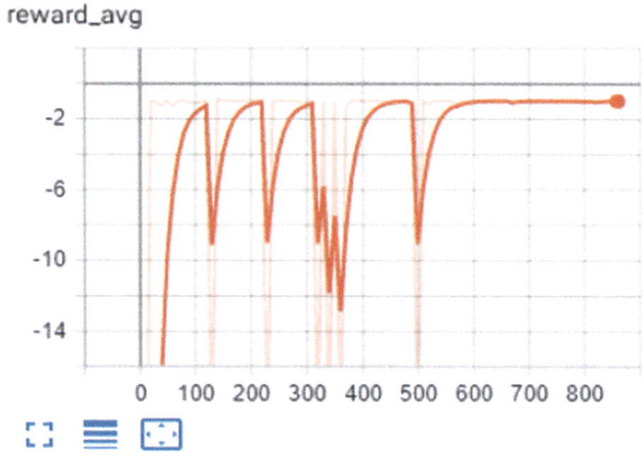

Fig. 10 Xception model training for 850 episodes

Model Performance in Different Weather Conditions

It was observed that during the training process, the model also performed well even when the weather presets were changed to rainy, sunny, and cloudy. The training was done on a computer CPU with a relatively low computational power and the training was done for 3000 episodes.

The model accuracy showed some reduction in performance accuracy as it was observed there were more collisions under the same period of training. But kept on improving as the training was increased. This is an indicator that the model can work under different scenarios. The relative speed and distance travelled by the agent within the model increased after there was an increase in the training of the model. It is from this point that the model can be deployed in the environment under supervision as it learns more scenarios and be able to navigate within the environment autonomously.

5 Recommendations and Conclusions

5.1 Conclusions

In this research, a reinforcement learning model was developed. With this model, the vehicle gathers data using an RGB Camera and collision sensor while moving on the road in a simulated environment. Scenarios are developed which include car moving around corners, straight road, and in a more urban layout with other obstacles like cars within the environment. Reward flags are given for no collision and penalty for collision with obstacles within the environment. Model testing was done in CARLA's simulator and analysis of the model was done on a Tensor board and recorded simulation as the vehicles moved within the environment.

The results obtained demonstrate that the model learned to perform actions, that is, brake and move forward within the environment. It mimics human driving and the algorithm was able to control the vehicle on the road. It has been tested on straight roads, corners, and in an environment with complex scenarios with cyclists and other cars.

The algorithm was able to actuate the vehicle and drive well on a straight road while maintaining its lane without getting off the lane and also brakes in case of an imminent collision. The algorithm is able to determine corners with more training and can turn around them. An environment with vehicles and other actors like cyclists was modelled and training was done, it can be observed that the vehicle was able to stop when it met another vehicle or any other cyclist.

5.2 Recommendations

In the future in an attempt to commercialize the model for deployment, more training should be done on a high-end GPU and for a longer period of time as this will increase the accuracy of the model and be able to pick random actions from the environment and be able to predict from them while the driving process is ongoing.

Other environments different from Carla environment may be used to achieve how the algorithm works and be able to generalize for even semi-urban and rural areas. Since this specific model is customized on highways and urban areas as opposed to rural areas.

In future, more sensors may be used like Lidar for range measurement by ray casting and also Radar for detection of objects in range. This will help in achieving the most efficient algorithm for autonomous cars.

It is prudent to employ different techniques apart from reinforcement learning for example deep learning and compare the different algorithms' performance for decision-making purposes.

References

1. MacHer, G., Schmittner, C., Dobaj, J., Armengaud, E., & Messnarz, R. (2020, April). An integrated view on automotive SPICE, functional safety and cyber-security. In *SAE Technical Papers* (Vol. 2020). https://doi.org/10.4271/2020-01-0145.
2. Lajunen, T., & Sullman, M. J. M. (2021). Attitudes toward four levels of self-driving technology among elderly drivers. *Frontiers in Psychology, 12*, 682973. https://doi.org/10.3389/fpsyg.2021.682973
3. Mattas, K., et al. (2020). Fuzzy surrogate safety metrics for real-time assessment of rear-end collision risk. A study based on empirical observations. *Accident; Analysis and Prevention, 148*, 105794. https://doi.org/10.1016/j.aap.2020.105794
4. Sanjana, T., Wahid, F., Habib, M. M., & Rumel, A. A. (2018). Design of an automatic forward and back collision avoidance system for automobiles. *Advances in Science, Technology and Engineering Systems, 3*(1), 205–212. https://doi.org/10.25046/aj030124
5. Koojo, I., Machuve, D., Mirau, S., & Miyingo, S. P. (2021, September). Design of a passenger security and safety system for the Kayoola EVs Bus. In *IEEE AFRICON Conference* (Vol. 2021). https://doi.org/10.1109/AFRICON51333.2021.9570932.
6. GitHub - AdoniaMbarebaki/Collision-avoidance-system-in-carla-v-0.9.5. Retrieved September 10, 2022, from https://github.com/AdoniaMbarebaki/Collision-avoidance-system-in-carla-v-0.9.5
7. Yuan, C., Wang, T., Shen, J., He, Y., & Weng, S. (2021). A new active safety distance model of autonomous vehicle based on sensor occluded scenes. *International Journal of Modelling and Simulation, 41*(3), 195–208. https://doi.org/10.1080/02286203.2019.1706068
8. Chae, H., & Yi, K. (2020). Virtual target-based overtaking decision, motion planning, and control of autonomous vehicles. *IEEE Access, 8*, 51363–51376. https://doi.org/10.1109/ACCESS.2020.2980391
9. Hosny, A., Yousef, M., Gamil, W., Adel, M., Mostafa, H., & Darweesh, M. S. (2019). Demonstration of forward collision avoidance algorithm based on V2V communication. https://doi.org/10.1109/MOCAST.2019.8741580.

10. Yousef, M., et al. (2019, August). Dual-mode forward collision avoidance algorithm based on vehicle-to-vehicle (V2V) communication. In *Midwest Symposium on Circuits and Systems* (Vol. 2018). https://doi.org/10.1109/MWSCAS.2018.8623896.
11. Yang, J., Kim, S., & Huh, K. (2020). Development of a unified lane-keeping and collision avoidance system for semi-trailer truck. *IEEE Access, 8*, 149751–149763. https://doi.org/10.1109/ACCESS.2020.3016497
12. Liu, L. C., Fang, C. Y., & Chen, S. W. (2017). A novel distance estimation method leading a forward collision avoidance assist system for vehicles on highways. *IEEE Transactions on Intelligent Transportation Systems, 18*(4), 937–949. https://doi.org/10.1109/TITS.2016.2597299
13. Aradi, S. (2022). Survey of deep reinforcement learning for motion planning of autonomous vehicles. *IEEE Transactions on Intelligent Transportation Systems, 23*(2), 740–759. https://doi.org/10.1109/TITS.2020.3024655
14. Rill, R. A., & Faragó, K. B. (2021). Collision avoidance using deep learning-based monocular vision. *SN Computer Science, 2*(5), 375. https://doi.org/10.1007/s42979-021-00759-6
15. Kilicarslan, M., & Zheng, J. Y. (2019). Predict vehicle collision by TTC from motion using a single video camera. *IEEE Transactions on Intelligent Transportation Systems, 20*(2), 522–533. https://doi.org/10.1109/TITS.2018.2819827
16. Wulfe, B., Hartong-Redden, R., Chintakindi, S., Kodali, A., Choi, S. C. T., & Kochenderfer, M. J. (2018). Real-time prediction of intermediate-horizon automotive collision risk. In *Proceedings of the International Joint Conference on Autonomous Agents and Multiagent Systems,* AAMAS (Vol. 2).
17. Venkateswaran, N., Hans, W. J., & Padmapriya, N. (2021). Deep learning based robust forward collision warning system with range prediction. *Multimedia Tools and Applications, 80*(14), 20849–20867. https://doi.org/10.1007/s11042-021-10703-8
18. Sakic, N., Krunic, M., Stevic, S., & Dragojevic, M. (2020, November). Camera-LIDAR object detection and distance estimation with application in collision avoidance system. In *IEEE international conference on consumer electronics – Berlin (Vol. 2020)*. ICCE-Berlin. https://doi.org/10.1109/ICCE-Berlin50680.2020.9352201
19. Shi, C., Dong, Z., Pundlik, S., & Luo, G. (2019). A hardware-friendly optical flow-based time-to-collision estimation algorithm. *Sensors (Switzerland), 19*(4), 807. https://doi.org/10.3390/s19040807
20. del Egido Sierra, J., Díaz, A., Bergasa, L. M., Barea, R., & López, M. E. (2021). Autonomous vehicle control in CARLA challenge. *Transportation Research Procedia, 58*, 69–74. https://doi.org/10.1016/j.trpro.2021.11.010
21. Malik, S., Khan, M. A., and El-Sayed, H. (2021). CARLA: Car Learning to Act - An inside out. In *Procedia Computer Science* (Vol. 198). 10.1016/j.procs.2021.12.316.
22. GitHub-carla-simulator: Reinforcement learning baseline agent trained with the Actor-critic algorithm. Retrieved September 10, 2022, from https://github.com/carla-simulator/reinforcement-learning
23. CARLA Simulator. Retrieved September 10, 2022, from https://carla.readthedocs.io/en/latest/
24. Marina, L., & Sandu, A. (2017). Deep reinforcement learning for autonomous vehicles-state of the art. *Bulletin of the Transilvania University of Braşov, 10*(59), 195–202.
25. Pérez-Gil, Ó., et al. (2022). Deep reinforcement learning based control for autonomous vehicles in CARLA. *Multimedia Tools and Applications, 81*(3), 3553–3576. https://doi.org/10.1007/s11042-021-11437-3
26. Kim, J., & Cho, J. (2018). On the reinforced reliability of forward collision warning system with machine learning. *International Journal of Mechanical Engineering and Technology, 9*(5), 1058–1063.

Visualizing Outlier Explanations for Mixed-Type Data

Jakob Nonnenmacher and Jorge Marx Gómez

1 Introduction

Outlier explanation approaches are used to explain the results of oftentimes untransparent and not intuitively understandable outlier detection methods. Outlier detection methods are used in different fields from accounting to fraud detection to cybersecurity. However, detecting the outlier is only the first step and turning the detected outlier into an actionable insight is the key challenge. One reason for this is that outlier detection methods are often employed in a more exploratory context thus requiring the analyst to further investigate the results before they are able to take action. Another reason is that outlier detection methods are good at providing information about which entries in a dataset are outlying—using a label or a score—but they rarely provide information for why an entry is outlying [1, 2]. This is especially true for newer but at the same time more performant approaches like deep learning or ensemble methods [3]. Finally, all outlier detection approaches utilize some kind of statistical measure to determine the outliers in a dataset. These statistical measures can select entries as outliers that are not seen as outliers according to the domain the detector is used in [4]. To filter out those potential false positives, the analyst must also understand why an entry has been selected as an outlier.

Outlier explanation approaches are especially relevant when the dataset the detection is performed on is high-dimensional. The reason for this is that it can be easier to get an overview of all features in a low-dimensional dataset for the analyst. They could, for example, plot the data as two-dimensional scatterplots in which they could then more easily see why an entry has been selected as an outlier. Datasets

J. Nonnenmacher (✉) · J. Marx Gómez
University of Oldenburg, Oldenburg, Germany
e-mail: jakob.nonnenmacher@uol.de

J. Marx Gómez et al. (eds.), *Artificial Intelligence Tools and Applications in Embedded and Mobile Systems*, Progress in IS,
https://doi.org/10.1007/978-3-031-56576-2_14

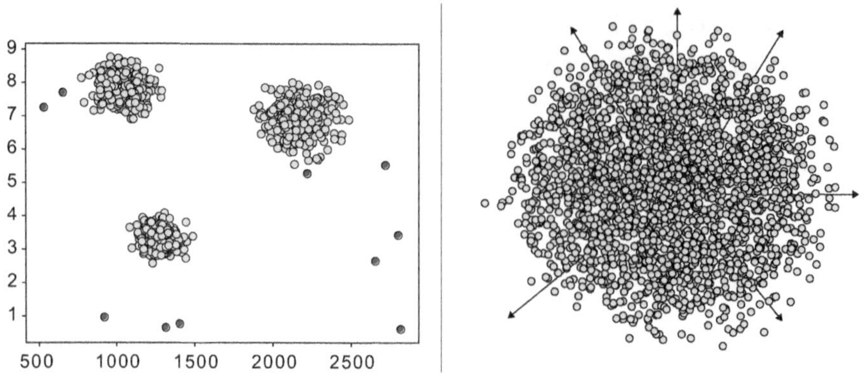

Fig. 1 Obvious outliers in 2D dataset and outliers hidden in an n-dimensional dataset

with higher dimensionality, this would not be feasible. For a dataset with only 20 dimensions, the number of possible plots would already be 190. An example of this difficulty can be seen in Fig. 1. Many real-world datasets are high-dimensional [5], with potential of hiding the relevant outlier causes.

Only two outlier explanation methods have focused on how to visualize the explanation. These are the approaches by [6, 7] which explain detected outliers using scatterplots. The approaches work well for enabling analysts to easily grasp and understand the causes of outliers for further analysis. These two approaches have been developed for numerical data which is the case for most outlier explanation approaches [8]. Only recently, approaches have been introduced to explain outliers detected in mixed-type data. This includes the approaches by [2] using isolation trees and [9] using attention-based neural networks. While these approaches tackle the important challenge of explaining outliers detected on mixed-type data—real-world data is often mixed-type [10]—they do not address how these explanations could be further improved using visualizations. What these approaches currently provide is a list of important features without further considering how they can be visualized.

For humans, it can be much easier to process information when they are visualized properly. The reason for this is that vision is a human's dominating sense [11] and its brain is specialized for processing information visually [12]. This is why visualizations are such an efficient way of presenting and analyzing data. Visualizations enable the exploitation of the human vision system for pattern searching when dealing with large and complex data [13].

For these reasons, we iteratively develop an approach for visualizing outlier explanations on mixed-type data in this paper. We then evaluate the approach in a focus group with internal auditors as potential users of outlier detection and outlier explanation. Internal auditing was selected for developing the approach since it presents a key group of users for outlier detection. Auditors are often faced with new topics for which they do not have experience yet [14]. At the same time, they are confronted with a growing amount of large and complex datasets which are often mixed-type. Overall, the contribution of this work includes:

- Developing a new outlier visualization approach for mixed-type data.
- Evaluation of this approach with potential users.
- Recommendations for avenues for further research.

The remainder of this paper is organized as such: Section 2 introduces related work. Section 3 presents our developed approach. Section 4 contains the evaluation as well as the discussion of our results and in Sect. 5, we conclude our work and provide an outlook for further research.

2 State-of-the-Art

Our research is positioned at the intersection of outlier explanation and visualization to support the utilization of outlier detection in different domains in which mixed-type data is prevalent. For this reason, we provide an overview of both outlier explanation techniques as well as methods to visualize mixed-type data.

2.1 *Outlier Explanation*

A general overview of outlier explanation methods is provided in a recent review by [8]. This review points out the limited number of studies that address visualizations as a possible aid in the context of outlier explanation. It only describes two approaches that focus on visualizations for outlier explanation methods [6, 7]. Both of these approaches have been designed for numerical data. The approaches try to determine a number of 2D subspaces which can be used to explain multiple outliers at once. The subspaces are determined by scoring all possible 2D subspaces using an outlier detection method. In the case of their studies, an isolation forest method is used for scoring. Those subspaces in which multiple outliers receive the highest outlier score are then chosen as the explaining subspace. Finally, the 2D subspaces are visualized in scatterplots. These approaches work well for the kinds of datasets they have been designed for. One downside their approaches have is that they will not be able to easily explain outliers which are outlying in subspaces of a dimensionality different from two.

Another observation the review by [8] makes, is that there is a limited number of methods that are able to explain outliers detected on mixed-type data. The methods able to provide an explanation for mixed-type outliers are the approaches by [2, 9]. The approach by [2] utilizes an isolation-tree method while the approach by [9] utilizes a neural network with self-attention. Both of these approaches provide information on which features are important for explaining an outlier but they do not offer visualizations for their explanations.

2.2 *Visualizations for Mixed-Type Data*

Most available data visualization methods have been designed for numerical data [13]. Only a few approaches exist for visualizing mixed-type data. One approach is the embedding of mixed-type data into lower dimensional numerical featurespaces to then plot it as numerical data [15]. The advantage of embedding the data as numerical data is that any visualization method that is suitable for numerical data can subsequently be used to visualize the data. These approaches have the potential to correctly translate the distances between the different entries in the dataset to the lower dimensional featurespace. This way, it can be seen that something is an outlier. One potential downside is that they do not show the original features which can still make it difficult to understand why something is an outlier.

[16] describes parallel coordinate plots as a possible method for visualizing multiple features of mixed-type datasets at the same time. One disadvantage they see for parallel coordinate plots is so-called swamping. In swamping, the entries which could be interesting to the analyst are covered by other less interesting entries in the plot [16]. The approaches by [10, 17] recommend adjustments to parallel coordinate plots to improve the ordering of categorical values within them and to improve the representation of correlations.

Swarm plots are another way of visualizing mixed-type data. They are an alternative to scatterplots when the goal is to visualize one numerical and one categorical variable in a plot [18]. A downside is that it can only show two-dimensional relationships and in large datasets points can overlap.

3 Research Setting and Method

We developed our method in two iterations and used the results from both iterations to adapt and improve our methods. For each iteration, we applied our approach within an internal audit. As the input for our method, an arbitrary outlier detection method in combination with an arbitrary outlier explanation method which provides importance values for each feature can be used. The specific detection method we used in our research to obtain the outliers has been a Trinity method as described by [5] and the specific explanation method to obtain the feature importance was the MIXATON approach described by [9].

Since outliers can have their cause in multiple subspaces, we decided to develop our outlier visualization approach based on a method which can be utilized to show multiple features at once. We also wanted the analyst to be able to see the original feature values. To enable this, we utilized a table in our prototype for the first iteration in which we highlighted the important features based on the obtained

	sex	length	diameter	height	whole weight	shucked weight	viscera_weight	shell weight	outlierscore
942	F	0,455	0,355	1,13	0,594	0,332	0,116	0,1335	25,19852041
583	M	0,705	0,565	0,515	2,21	1,1075	0,4865	0,512	8,744045376
1798	I	0,725	0,55	0,22	2,0495	0,7735	0,4405	0,655	4,926027511
170	M	0,155	0,11	0,04	0,0155	0,0065	0,003	0,005	3,085013105
1265	M	0,725	0,57	0,19	2,3305	1,253	0,541	0,52	3,083761434
1147	F	0,705	0,56	0,205	2,381	0,9915	0,5005	0,624	3,065600874
97	I	0,11	0,09	0,03	0,008	0,0025	0,002	0,003	2,893393511
584	F	0,745	0,57	0,215	2,25	1,1565	0,446	0,558	2,844307363
1204	I	0,405	0,31	0,11	0,91	0,416	0,2075	0,0995	2,830901633

Fig. 2 Highlighting of relevant features in the table showing the Abalone dataset

feature importance. An example of this using data from the mixed-type Abalone[1] dataset is provided in Fig. 2.

Applying the approach like this worked well but had two downsides. The first one was that the explanation showed which features were important together with its value for making an entry outlying but it did not show how that outlier differed from normal entries with similar values. Additionally, the detection and explanation uncovered a large number of outliers with the same underlying cause. Being able to group these outliers together could reduce the analyst's effort to individually check each entry. These shortcomings were addressed in the prototype for the second iteration.

To group outliers with similar causes together, we utilized the obtained feature importance values in combination with the feature values and the obtained outlier score. As a clustering method, we employed DBSCAN since it does not require the number of clusters to be predetermined by the analyst. This way, a number of different outlier groups are obtained for a dataset. In the first part of the prototype, the obtained entries with the highlighted importance and the groups are then shown to the analyst. The analyst can then select one of the groups to have it explained in more detail. For this, a parallel coordinate plot is used. This plot then also allows to show how an outlier group or entry differs from inlier entries.

To avoid the problem of swamping in the parallel coordinate plot and to incorporate the information provided by the feature importance values, we adapted the plot in the following way: The cluster selected by the analyst was color coded in red, the remaining outlier clusters were color coded in yellow, and the normal entries were color coded in blue. All entries of the selected cluster are shown, one entry for each other outlier cluster is shown and a random sample of 10,000 normal entries is shown. Additionally, the axes in the plot are sorted based on the importance values determined by the explanation method. We combined the plot with a table that shows the entries of the dataset underneath it. In the table, the important attributes per entry are highlighted. Finally, we allowed for interactive filtering on multiple axes which

[1] https://archive.ics.uci.edu/ml/datasets/abalone.

at the same time filtered the entries in the table. This adapted version was then applied in a second iteration on another audit which yielded good results.

An example of the unaltered parallel coordinate plot and the parallel coordinate plot with our adaptions is provided in Figs. 3 and 4 on the mixed-type Abalone dataset. The dataset has been chosen since larger dimensional datasets would be

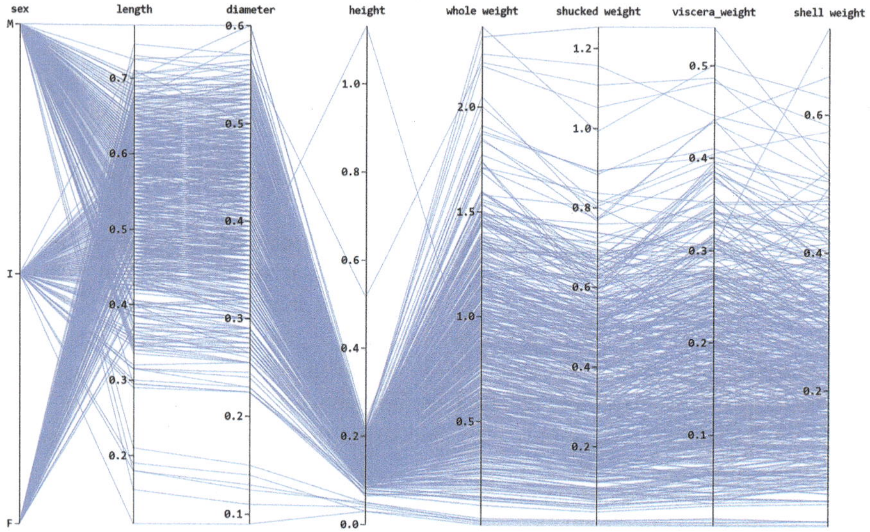

Fig. 3 Unaltered parallel coordinate plot showing the Abalone dataset

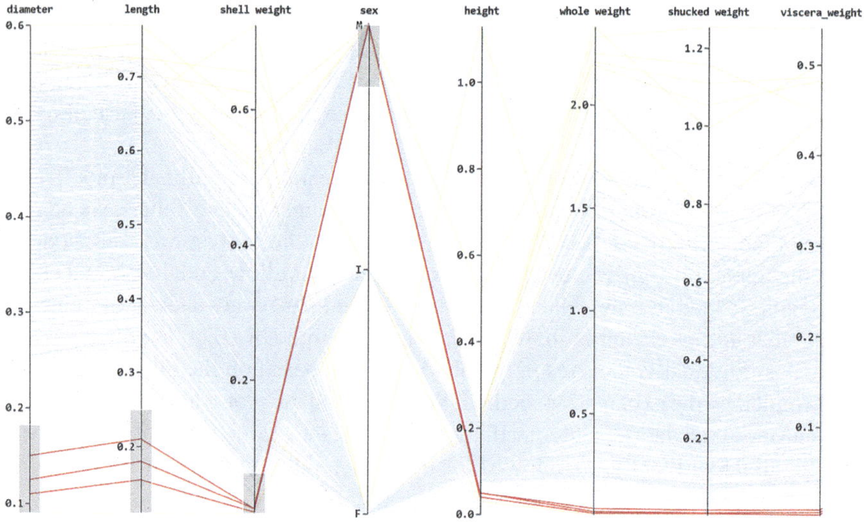

Fig. 4 Adapted parallel coordinate plot with a selected outlier cluster, sorted columns, and filtered on important features showing the Abalone dataset

difficult to fit into the margins of this paper. For this reason, the axis for the outlier score and outlier kind were also omitted in the plot. The prototype was implemented using Python and JavaScript. The parallel coordinate plot was implemented by adapting the d3.paarcoords library.[2]

4 Evaluation

After applying the approach in a second iteration, a focus group [19] with auditors from an international automotive manufacturer was performed for a structured evaluation. The goal of this evaluation was to gather feedback from the auditors regarding the current version of the prototype as well as to gather ideas for further adjustments and improvements.

The workshop was conducted with three auditors. For the purpose of the workshop, the prototype was applied to data from two audits the auditors were currently conducting. To structure the workshop, a number of guiding questions were used. Based on these guiding questions, the discussion with the auditors was facilitated. When necessary, different aspects of the prototype were shown to the auditors. The feedback that was received from the auditors together with the guiding questions is shown in a summarized form in Table 1.

Based on the workshop the general usefulness of the developed explanation visualization approach is shown. The parallel coordinate plot with our adaptions can clearly show the difference between outliers and inliers and can thus help the analyst in further investigations of the outliers. The recommendations by the workshop participants also show how the visualizations could be further improved. This includes adaptions which could be made to facilitate the communication with business users necessary for examining or confirming the outliers.

5 Discussion and Conclusion

We present a first approach for enabling the visualization of mixed-type outlier explanations. Our approach works with arbitrary outlier detection and explanation methods given they provide the feature importance of outliers. We developed and evaluated this approach in a real-world environment with great potential for outlier detection and explanation. Due to its general nature, the approach could also facilitate the explanation of outliers in other domains that utilize mixed-type data.

One possible avenue for further research would be the combination of the developed approach with further visualization approaches depending on the dimensionality of the explaining subspace. If the feature importance returned for an

[2] https://syntagmatic.github.io/parallel-coordinates/.

Table 1 Results of focus group

Question	Auditors' feedback
1. Does the visualization and explanations improve the understanding of the outliers?	• Generally useful • Could see themselves working with it • Like how it shows the contrast between inliers and outliers • Like the ability to filter on explaining features • Like that one could see how many other entries had similar values in accompanying table
2. Does the visualization provide enough information?	• Opinions differed • No: It should be more apparent that not all entries are shown in datasets of more than 10,000 entries • Yes: Cluster shown in full is what important
3. If not, what kind of visualization / information is missing?	• The additional written explanation would be nice • Show frequency distributions for each feature as additional information
4. How could the visualization be improved?	• Almost too much information • Do not show all features in the plot but only the important ones • Group / cluster the normal entries as well to not have to show all of them • Hide irrelevant columns (based on feature importance or based on user selection) • Features with only one value should be excluded from plot • Sort features by variance
5. Do you trust the explanation of the outliers?	• Partially, the sampling for visualization in large datasets makes it difficult • More information is necessary
6. Would this explanation help you in utilizing outlier detection?	• Yes, especially thanks to the filtering • Maybe have the parallel coordinate plot first to then drill down to the table • The interactive plot could be used directly in meetings with business users • Ideally, there should be a way to define an "ID" column that can then be copied to directly provide it to the auditee for checking in their system • Download selected entries as excel file to provide to auditee when synchronized meeting is not possible • Ability for exporting the plot would be interesting as well

outlier clearly indicates that an outlier is only outlying in a one- or two-dimensional featurespace, appropriate visualization methods for these smaller subspaces could be combined with the parallel coordinate plot. In the future, the approach could be evaluated in further domains including mobile or embedded systems.

References

1. Liu, N., Shin, D., & Hu, X. (2018). Contextual outlier interpretation Proceedings of the 27th International Joint Conference on Artificial Intelligence (pp. 2461–2467). AAAI Press, Stockholm, Sweden.
2. Kopp, M., Pevný, T., & Holeňa, M. (2020). Anomaly explanation with random forests. *Expert Systems with Applications, 149*, 1–18. https://doi.org/10.1016/j.eswa.2020.113187
3. Carletti, M., Terzi, M., & Susto, G. A. (2020). Interpretable anomaly detection with DIFFI: Depth-based feature importance for the isolation Forest.
4. Siddiqui, M. A., Stokes, J. W., Seifert, C., Argyle, E., McCann, R., Neil, J., & Carroll, J. (2019). Detecting cyber attacks using anomaly detection with explanations and expert feedback. In *ICASSP 2019–2019 IEEE International Conference on Acoustics, Speech and Signal Processing (ICASSP)* (pp. 2872–2876). https://doi.org/10.1109/ICASSP.2019.8683212
5. Aggarwal, C. C. (2017). *Outlier analysis*. Springer.
6. Gupta, N., Eswaran, D., Shah, N., Akoglu, L., & Faloutsos, C. (2018). Beyond outlier detection: LOOKOUT for pictorial explanation. In *ECML PKDD*.
7. Liu, H., Ma, F., Wang, Y., He, S., Chen, J., & Gao, J. (2020). LP-explain: Local pictorial explanation for outliers. In *2020 IEEE International Conference on Data Mining (ICDM)*. https://doi.org/10.1109/ICDM50108.2020.00046
8. Nonnenmacher, J., Holte, N., & Marx Gómez, J. (2022). Tell me why - A systematic literature review on outlier explanation for tabular data. In *3rd International Conference on Pattern Recognition and Machine Learning*.
9. Nonnenmacher, J., & Gómez, J. M. (2022). Beyond numerical–MIXATON for outlier explanation on mixed-type data. In *The 34th International Conference on Software Engineering & Knowledge Engineering*. https://doi.org/10.18293/SEKE2022-110
10. Zhang, Z., McDonnell, K. T., Zadok, E., & Mueller, K. (2015). Visual correlation analysis of numerical and categorical data on the correlation map. *IEEE Transactions on Visualization and Computer Graphics, 21*, 289–303. https://doi.org/10.1109/TVCG.2014.2350494
11. Hutmacher, F. (2019). Why is there so much more research on vision than on any other sensory modality? *Frontiers in Psychology, 10*, 481030.
12. Alawadhi, A. (2015). The application of data visualization in auditing. https://doi.org/10.7282/T3GQ70MD.
13. Liu, S., Maljovec, D., Wang, B., Bremer, P.-T., & Pascucci, V. (2017). Visualizing high-dimensional data: Advances in the past decade. *IEEE Transactions on Visualization and Computer Graphics, 23*, 1249–1268. https://doi.org/10.1109/TVCG.2016.2640960
14. Nguyen, L., & Kohda, Y. (2017). Toward a model of wisdom determinants in the auditing profession. In *Proceedings of the 50th Hawaii International Conference on System Sciences* (pp. 4604–4613). https://doi.org/10.24251/HICSS.2017.560
15. Grané, A., & Romera, R. (2016). On visualizing mixed-type data: A joint metric approach to profile construction and outlier detection. *Sociological Methods & Research, 47*, 207–239. https://doi.org/10.1177/0049124115621334
16. Wilkinson, L. (2018). Visualizing big data outliers through distributed aggregation. *IEEE Transactions on Visualization and Computer Graphics, 24*, 256–266. https://doi.org/10.1109/TVCG.2017.2744685
17. Rosario, G. E., Rundensteiner, E. A., Brown, D. C., Ward, M. O., & Huang, S. (2004). Mapping nominal values to numbers for effective visualization. *Information Visualization, 3*, 80–95. https://doi.org/10.1057/palgrave.ivs.9500072
18. Idrus, Z., Rusli, F. S., Nazri, M. A. M., Al-zebari, A., & Talib, N. H. A. (2021). Swarm Plot: Data Redistribution in Non-Random Technique. In *2021 6th IEEE International Conference on Recent Advances and Innovations in Engineering (ICRAIE)* (pp. 1–6). https://doi.org/10.1109/ICRAIE52900.2021.9703994
19. Tremblay, M. C., Hevner, A. R., & Berndt, D. J. (2010). The use of focus groups in design science research. In A. Hevner & S. Chatterjee (Eds.), *Design research in information systems: Theory and practice* (pp. 121–143). Springer US. https://doi.org/10.1007/978-1-4419-5653-8_10

A Dolutegravir-Associated Hyperglycemia Computational Prediction Tool for People Living with HIV in Uganda

Ceaser Wisdom Favor, Sinde Ramadhan, Michael Kisangiri, Levicatus Mugenyi, Francis Musinguzi, Martin Balaba, Noela Owarwo, Eva Laker, Ruth Obaikol, Agnes Kiraga, Barbara Castelnuovo, and Rosalind Parkes-Ratanshi

1 Introduction

Dolutegravir (DTG) is a second-generation integrase strand transfer inhibitor (INSTI) approved by the US Food and Drug Administration (FDA), and recommended as a first-line HIV treatment globally by the World Health Organisation (WHO) because of its efficacy and safety [1–5]. DTG has offered a critical

An Innovation Paper for Presentation at the First International Conference on Technological Advancement in Embedded and Mobile Systems (ICTA-EMoS) November 24–25, 2022; Arusha, Tanzania. Theme 2: Artificial Intelligence Tools and Applications: Machine Learning Applications (ML).

C. W. Favor (✉)
School of Computational and Communication Science and Engineering, Nelson Mandela African Institution of Science and Technology, Arusha, Tanzania

Infectious Disease Institute, Kampala, Uganda
e-mail: favorw@nm-aist.ac.tz; favorw@nm-aist.ac.ug; wceasar@idi.co.ug

S. Ramadhan · M. Kisangiri
School of Computational and Communication Science and Engineering, Nelson Mandela African Institution of Science and Technology, Arusha, Tanzania

L. Mugenyi
The AIDS Support Organisation, Kampala, Uganda

F. Musinguzi · M. Balaba · N. Owarwo · E. Laker · R. Obaikol · A. Kiraga · B. Castelnuovo
Infectious Disease Institute, Kampala, Uganda

R. Parkes-Ratanshi
Infectious Disease Institute, Kampala, Uganda

Department of Medicine, University of Cambridge, Cambridge, UK

J. Marx Gómez et al. (eds.), *Artificial Intelligence Tools and Applications in Embedded and Mobile Systems*, Progress in IS,
https://doi.org/10.1007/978-3-031-56576-2_15

Antiretroviral treatment (ART) option of daily dosing, minimal toxicity, viral suppression, and a solution to drug resistance than nevirapine and efavirenz therapeutic options [6–9]. However, studies have shown the development of non-communicable diseases (NCDs) inter liar in PLWH as a side effect of ART [10–16]. In Sub-Saharan Africa, the prevalence of Diabetes among PLWH is placed at 5.8% points higher (95% confidence interval (CI) 1.8 pp–9.8 pp) compared to individuals without HIV and those on ART are 4.6% points higher (95% CI 2.6 pp–6.6 pp) [17]. Moreover, Uganda's prevalence of diabetes mellitus among PLWH is 4.7% (95% CI, 2.6–6.8) [18]. Furthermore, studies uncovered that switching treatment-experienced and/or naïve PLWH to DTG also presented challenges of DTG-associated hyperglycemia [19–27]. The development of hyperglycemia by PLWH on DTG-based ART in Uganda has been placed at a rate of 0.47% per year between April 1, 2018, and March 31, 2019 [28], though hyperglycemia is said to improve when PLWH are discontinued from dolutegravir [29]. The Uganda Ministry of Health implemented a manual hyperglycemia screening tool [30]. This is limited to only screening for already hyperglycemic or diabetic PLWH before being switched to DTG. Machine learning models have been developed for the prediction of hyperglycemia as an NCD [16, 31, 32], but there is no available research yet undertaken to use Machine Learning techniques in the prediction of DTG-Associated hyperglycemia as a comorbidity PLHIV. Therefore, this study aims to develop a machine learning prediction tool to screen treatment-experienced PLWH for the risk of developing hyperglycemia when switched to DTG.

2 Materials and Methods

2.1 Organization and Ethics

The investigation team constituted five physicians, two statisticians, three data scientists, and one software engineer. The investigators began the study in 2020 by obtaining ethical approval from The AIDS Support Organization (TASO) research ethics committee. It was followed by institutional administrative approvals from the study sites.

2.2 Study Sites

The study sites purposefully chosen included the Infectious Disease Institute (IDI), TASO, and Mild-may which provide ART services in areas of Kampala, Wakiso, Entebbe, Jinja, and Mukono among others within Uganda. Participants' data is therefore representative of the Ugandan population, and the provision of ART and monitoring services at these sites has always followed the WHO and/or MOH guidelines right from the 1990s. The baseline ART regimens included nucleoside reverse inhibitors (NRTI) in combination with non-nucleoside reverse inhibitors

(NNRTI) [33]; after which participants were switched to the DTG-based regimens [1]. Participants initiated on ART at baseline in a clinic have a monthly follow-up and drug refill appointment. Given their stability, they are medically reviewed by a doctor after every 3 months. During doctor visits; Viral-load, CD4-Count, blood pressure, biodata, baseline ART start date, DTG regimen date, ART regimen information, and random blood sugar (RBS) measurements are registered in the organizations' medical information systems while ensuring systematic data quality control. Their medical information systems are Uganda-EMR, Clinic-Master, and Integrated Clinic Enterprise Application (ICEA) for TASO, Mild-may, and IDI, respectively. The participants who had started on DTG-based ART as early as the third quarter of 2017 gave the investigation a dataset rich in diverse durations on ART. The existence of electronic centralized databases made secondary data collection easier and allowed the generation of new insights and longitudinal analysis.

2.3 Study Procedure

Datasets extracted from organizational databases included treatment-experienced participants with these criteria:

- Were on a DTG-based regimen with no prior history of diabetes or hyperglycemia.
- Had been on DTG for at least 3 months.
- Had been on ART for not less than 2.5 years.
- Were aged between 3.5 and 90 years.
- Were not on a second line of ART.
- And had at baseline, been on a regimen other than DTG.

Subjects were Excluded using these criteria:

- Were on DTG as their baseline regimen.
- Were on DTG but had no random blood sugar (RBS) data at all.
- Were on diabetic drugs before switching to DTG.
- Were having unknown diabetes status.
- Were on DTG but switched to other regimens without captured RBS data or other reason for the switch.
- Were on DTG second-line regimens.

2.4 Data Collection

An abstract data acquisition tool was designed by the researchers to collect longitudinal data variables which included clinical information, basic demographics, and ART History. The description of the variables is recorded in Table 1. The

Table 1 Materials used for this study

Data processing, analysis, model, and tool development environments	Types of variables	Variable names	Description	Datatype
	1. Art history.	Previous/baseline regimen	Regimen at baseline or regimen used for the longest time	String
		Regimen-start-date	Date started on the first regimen	Date
		DTG regiment start date	The date started on the DTG regimen	Date
		Current regimen	DTG regimen switched to	String
	Clinical information	Viral-load	Most viral-load within 6 months before switching to DTG	Integer
		CD4-count	Most recent CD4-count within 6 months before switching to DTG	Integer
		TB-status	Most recent TB-status within 6 months before switching to DTG	String
		Pregnancy	Most recent pregnancy status within 6 months before switching to DTG	String
		RBS	Most recent RBS within 6 months before switching to DTG	Integer
			Also, the most recent RBS after switching to DTG	Integer
		Systolic pressure	Most recent systolic pressure within 6 months before switching to DTG	Integer
		Diastolic pressure	Most recent diastolic pressure within 6 months before switching to DTG	Integer

	Basic demographics	Gender		String
		Date of birth		Date
		Weight	Most recent weight within 1 year before switching to DTG	Integer
		Height	Most recent height within 1 year before switching to DTG	Integer
Excel sheet tool			In CSV format, for data preparation	
Anaconda environment			For data processing, analysis, and model building	
Programing language used			Python, XHML	
Atom development environment			For prediction tool development	

participants' total records for persons on DTG collected from each organization were: 57471, 40950, and 67645 for IDI, TASO, and Mild-may, respectively.

2.5 Data Processing

The datasets were processed using the Microsoft Excel tool in comma-supported values (CVS) format and the Anaconda machine learning development environment.

Participants' most recent values of variables; weight, height, viral load, CD4 count, systolic pressure, and diastolic pressure within 9 months prior to switching to DTG were extracted from the datasets.

Hyperglycemia was defined as an RBS of ≥11 mmol/L. Regimen duration was derived from the subtraction of baseline regimen start date from DTG regimen start date. Age was calculated by subtracting birth from current date. Body mass index (bminorm) was derived by dividing participants' body weights by their heights in meters squared.

This was followed by extraction of the TB-status, gender, baseline, or any other first-line regimen with the longest duration date, and the most recent RBS while on DTG.

After applying the study inclusion and exclusion criteria and machine learning-oriented data cleaning and processing, a clean participants dataset of 9077 was generated. Categorical variables were then nominal/ordinal encoded and null values were imputed using the Multivariate Imputation by Chained Equation (MICE) algorithm [34].

2.6 Model Building

Data was split in a ratio of 2:1; training and testing datasets respectively with certainty that hyperglycemia positive and negative participants are split in equal proportions into the two datasets. The medical data classification imbalance phenomena in the training dataset was solved using the Synthetic Minority Over-Sampling Technique (SMOTE) algorithm [35] thus addressing the potential bias of the model learning more of the majority class at the expense of the minority. Eight classifier algorithms were selected (Extreme Gradient Boost Classifier (XGB), Random Forest Classifier (RF), Support Vector Machine (SVM), Gaussian Naïve Bayes (NB), Decision Tree Classifier (CART), KN-Neighbors (KNN), Linear Discriminant Analysis (LDA), Linear Regression (LR)) basing on the causal characteristics and distribution of the variables' dataset. K-NN algorithm learns the similarity among observations/events and available classes and then classifies the new event into a class most similar within the available classes. Ensemble algorithms (RF, XGB) conglomerate several base models to yield an optimal predictive model. Tree-based models (CART, LDA) use a sequence of if-then rules to generate predictions from one or a number of decision trees. SVM uses a hyperplane to create classes into which events or observations are classified. NB uses a probability density function to classify events by taking new input variables and outputting the probability of belonging to a given class.

All the above different classifiers were run over the training dataset for ten iterations Each model's accuracy variations were plotted using a box and whisker comparison graph. Two (XGB and RF) of the eight selected models exhibited higher model performance accuracies.

The test dataset was used to evaluate and select the predictive model with the best classification metrics of: accuracy, sensitivity, positive predictive value, negative predictive value, F1-score, specificity, and area under a precision-recall curve.

The best model learning pattern was assessed using the hyperparameter and cross-validation score curves plotted on a graph of cross-validation scores against training instances.

To score the positive class, the discriminant threshold on the discriminant analysis graph was set to 0.45. In order to calibrate the model a plotted graph of fractions of positives against mean prediction probability was constructed to make sure the model replicates a reasonably good perfect estimation (prediction). Feature importance was conducted to determine the contribution of every variable to the model prediction. An optimal number of features that can contribute to the model prediction accuracy without creating any prediction redundancies.

The tool was developed using Python programming language and flask micro web framework.

3 Results

The results of model accuracies below show RF and XGB models at 0.988 and 0.989 which are higher than all accuracies the other models developed, Fig. 1. This called for further investigation of the consistency of model accuracies when subjected to the same dataset over a number of iterations.

```
('LR', 0.6770543277822008, 0.03305775709051789)
('LDA', 0.6949211881611664, 0.01685084404009784 2)
('KNN', 0.8553654074523841, 0.01577575769077790 5)
('CART', 0.8940641369278955, 0.007089790093361343)
('NB', 0.6122583150275729, 0.02107359817447 37)
('SVM', 0.5290422857790226, 0.02095709708902302)
('RF', 0.9881160399056702, 0.002196151286516070 3)
('XGB', 0.9891230537507869, 0.002245744820137256)
```

Fig. 1 Model accuracies for the eight ran algorithms

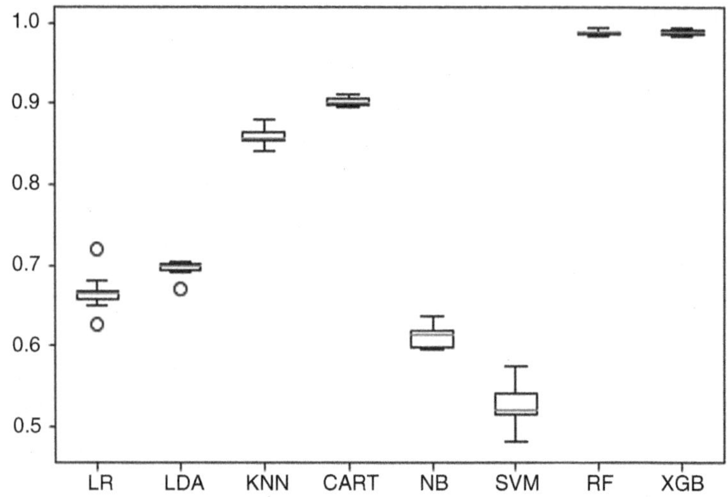

Fig. 2 Box and whiskers plot illustrations of accuracy variations for each of the models after ten iterations over the training dataset

3.1 Evaluation Measures

When each of the eight algorithms was subjected to ten iterations over the same training dataset, RF and XGB models had the least accuracy variations compared to all the other models. This meant the accuracy of these models was dependable. See Fig. 2.

3.2 Classification Report

The two better performing models (XGB, RF) classification metrics were evaluated using the testing dataset which divulged positive predictive values (PPV) (0.67, 0.60), negative predictive values (NPV) (0.98, 0.97), specificity (093, 0.92), Sensitivity (0.89, 0.83), F1-score of the negative class (0.96, 0.94), F1-score of the positive class (0.77, 0.70), and classification accuracy (0.93, 0.91), respectively, Figs. 3 and 4. The XGB model had a PPV of 0.07, NPV of 0.01, specificity of 0.01, sensitivity of 0.06, negative class F1-score of 0.02, and positive class F1-score of 0.07 higher than that of the RF model.

	precision	recall	f1-score	support
0.0	0.98	0.93	0.96	1176
1.0	0.67	0.89	0.77	186
accuracy			0.93	1362
macro avg	0.83	0.91	0.86	1362
weighted avg	0.94	0.93	0.93	1362

Fig. 3 XGB classification report

	precision	recall	f1-score	support
0.0	0.97	0.92	0.94	1184
1.0	0.60	0.83	0.70	178
accuracy			0.91	1362
macro avg	0.79	0.87	0.82	1362
weighted avg	0.92	0.91	0.91	1362

Fig. 4 RF classification report

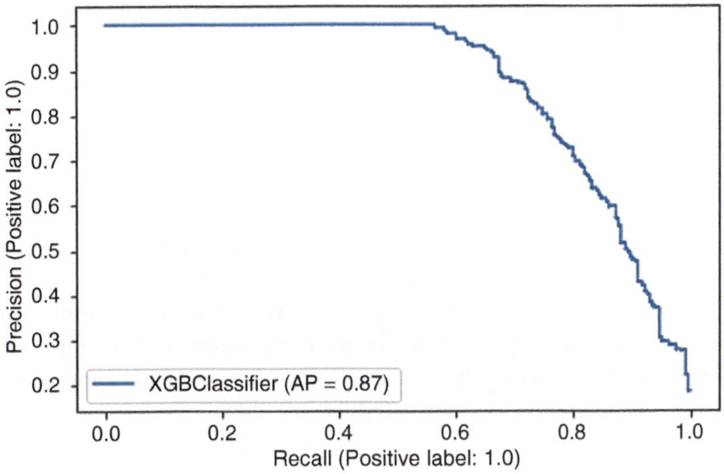

Fig. 5 XGB precision-recall curve

3.3 Precision-Recall Curve

A precision-recall curve is plotted by precision (PPV) against recall true positive rate (TPR), thus area under a precision-recall curve is a measure of the number of true positive classifications made. The XGB and RF area under precision curves are 0.87 and 0.81, respectively, Figs. 5 and 6. This implies the XGB model has a

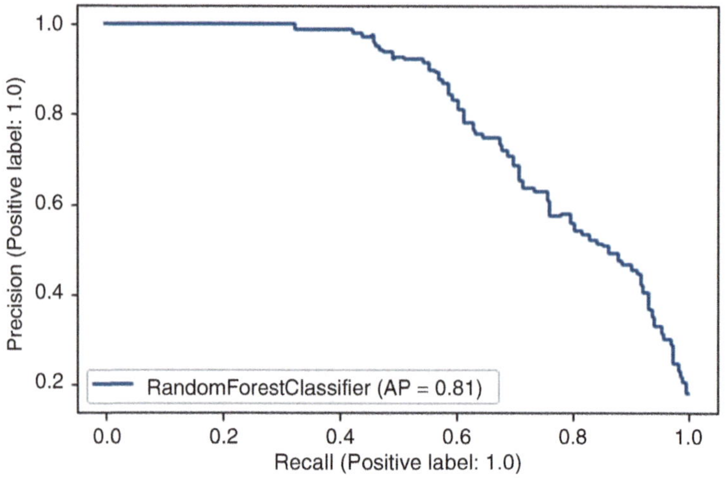

Fig. 6 RF precision-recall curve

0.06 higher predictive power for people who would truly develop hyperglycemia if switched to DTG than RF.

3.4 Feature Importance

Feature importance is aimed at improving model performance and simplicity. Applied model dimensionality reduction dropped the TB-Status feature and retained all the others as the important features for the model prediction. The plotted features results on the horizontal bar graph show ascending order contribution to the XGB model prediction, Fig. 7. Using the recursive feature elimination with cross-validation of random forest classifier for further feature assessments showed the model test score saturating after the eighth feature Fig. 8.

3.5 Hyperparameters and Model Validation

An XGB learning curve was a plot on the graph of cross-validation scores with respect to training instances. It exhibited convergence at the high scores with an increase in training instances. This suggested model improvement when subjected to more training instances, Fig. 9.

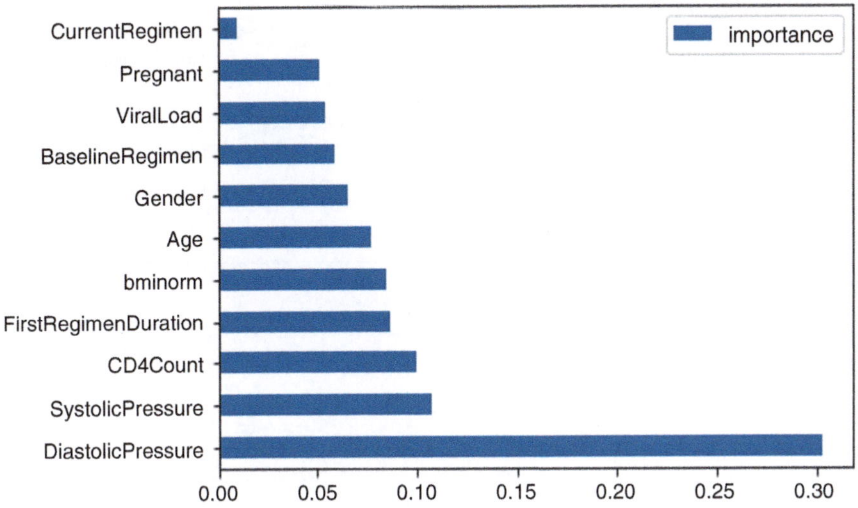

Fig. 7 Contributions of each feature model output

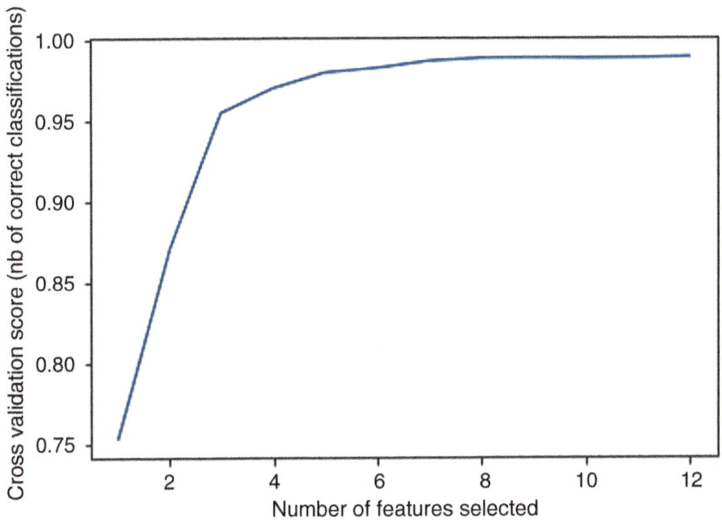

Fig. 8 XGB optimal number of features

3.6 Model Calibration

A calibration plot shows the XGB model (blue line) curve plotted to hang around the perfect estimation (black dotted line). Which exhibited the XGB model overestimating between (0.25–0.55) and (0.6–0.83) Fig. 10.

Mean Prediction Probability.

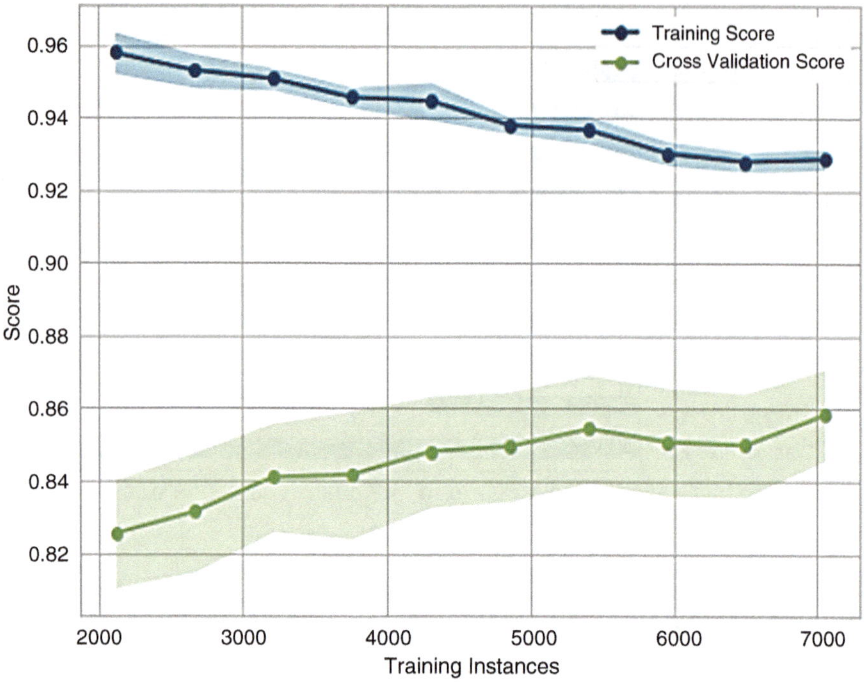

Fig. 9 Cross-validation of the XGB model learning

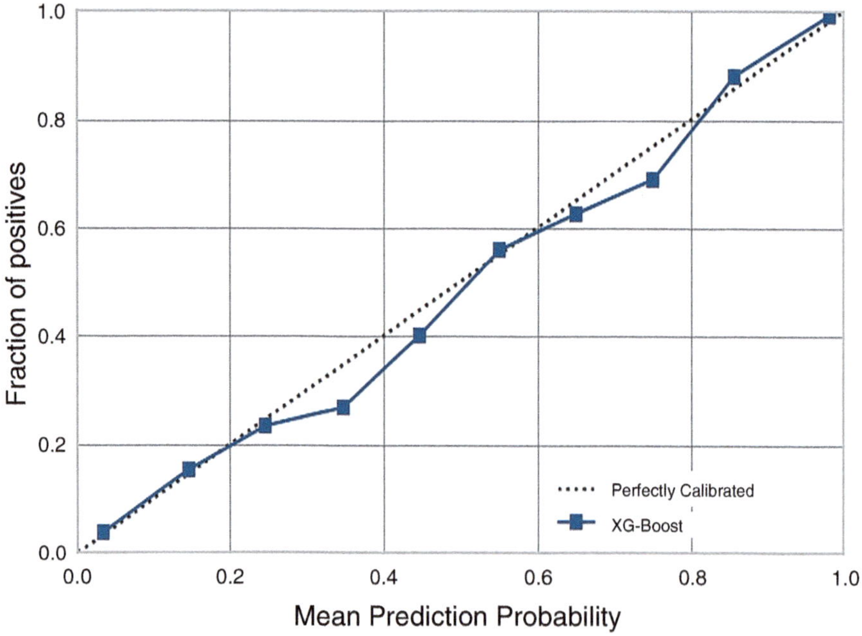

Fig. 10 XGB calibration plot (reliability curve)

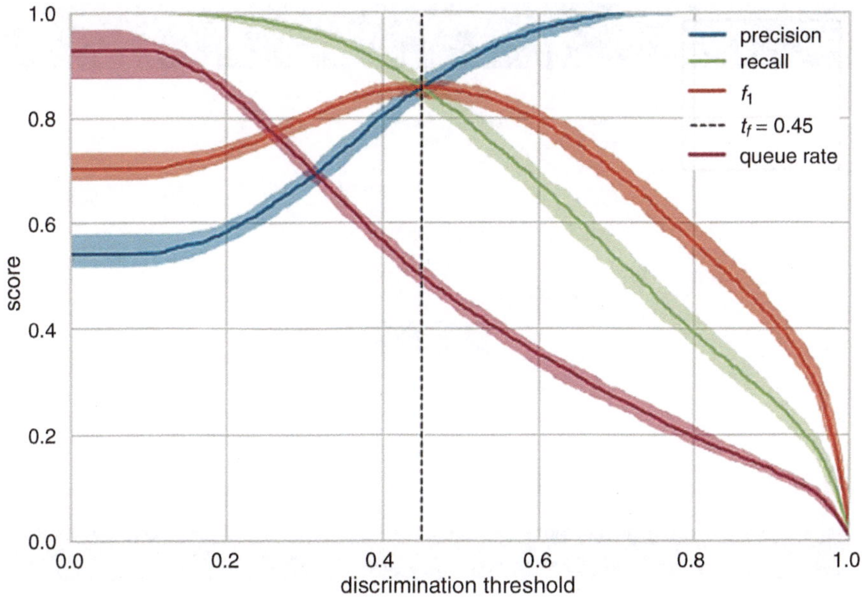

Fig. 11 The optimal discriminant threshold graph

3.7 *Optimal Discriminant Threshold*

Tuning of the discrimination threshold (0.45) was made with the consideration of metrics of Precision, Recall, and F1-score. Figure 11 shows interactions of Recall, Precision, and F1-score graphs interacting at the optimal threshold (0.84) score/probability where the XG-boost model was adjusted to increase sensitivity to false positives. Beyond this point, the recall and precision curves have an inverse relationship. And the model will predict more PLHIV to be true DTG-Associated hyperglycemia positive at a positive probability of 0.84 or above. This is because the recall curve probability score will decrease while the precision is increasing.

3.8 *DTG-Associated Hyperglycemia Prediction Tool*

The tool was developed with a web-based user interface which takes 11 features to make a prediction with the model. The results of the predictions are shown in Figs. 12 and 13.

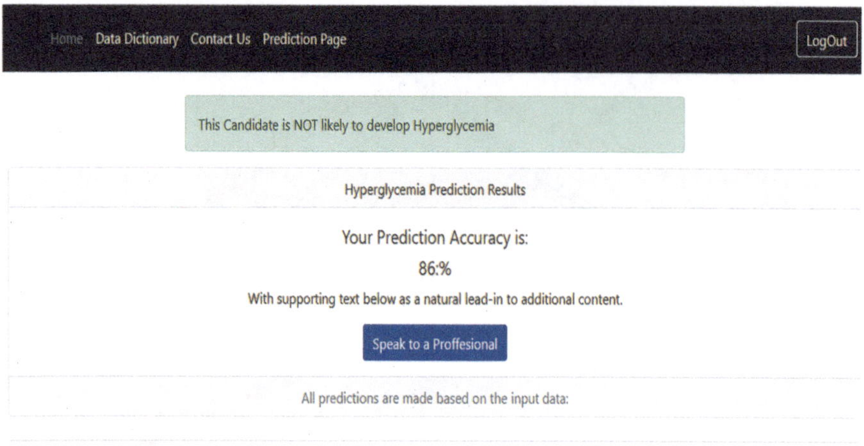

Fig. 12 Tools' negative classification

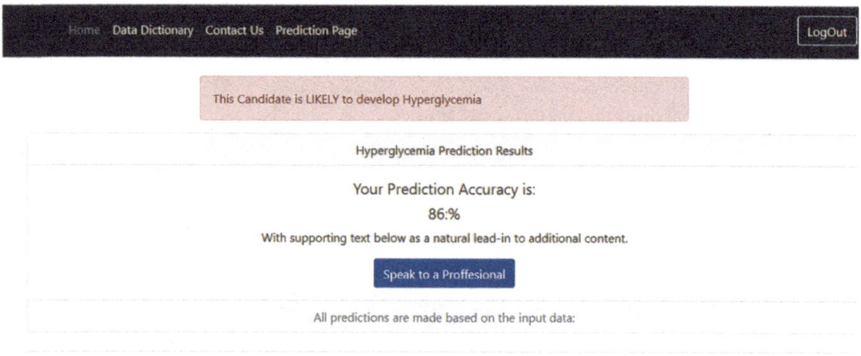

Fig. 13 Tool's positive prediction results

4 Discussions

Evaluation classification metrics implied the XGB model had a higher ability to classify treatment-experienced PLWH who would develop hyperglycemia as true positives than the RF.

The convergence at higher scores by the training and cross-validation score curves suggested that adding more learning instances will improve the XGB model and prediction accuracy.

XGB reliability graph illuminates a relatively fair estimation by the model. This infers reliable predictability of positives by the XGB model because the least pessimism or optimism tendencies were noticed in the XBG model in comparison to the perfect estimation.

5 Conclusion

From the deliberations of classification metrics and precision-recall results, the study suggests the XGB model as the most appropriate model for prediction tool development.

Most importantly the tool demonstrates that machine learning (ML) techniques can be applied to datasets of longitudinal cohorts of PLWH to help improve on positive ART outcomes.

6 Recommendations

The health domain should employ machine learning technics improving health. The DTG-associated hyperglycemia prediction tool should be used as a screening tool for HIV treatment-experienced patients being switched to DTG in Uganda. Given the longitudinal nature of the cohort data of PLWH, the study also recommends the use of time series ML algorithms to investigate the contribution of time to negative ART outcomes.

References

1. WHO. (2021). World Health Organisation (WHO). [Online]. Retrieved October 4, 2021, from https://apps.who.int/iris/bitstream/handle/10665/277395/WHO-CDS-HIV-18.51-eng.pdf?sequence=1&isAllowed=y
2. Dow, D. E., & Bartlett, J. A. (2014). Dolutegravir, the second-generation of integrase Strand transfer inhibitors (INSTIs) for the treatment of HIV. *Infectious Diseases and Therapy Springer Link, 3*(2), 83–102.
3. Scali, I., & Abernathy, A. (2021). [Online]. Retrieved October 4, 2021, from https://viivhealthcare.com/en-gb/media/press-releases
4. van Lunzen, J., Maggiolo, F., Arribas, J. R., Rakhmanova, A., Yeni, P., Young, B., Rockstroh, J. K., Almond, S., Song, I., Brothers, C., & Min, S. (2012). Once daily dolutegravir (S/GSK1349572) in combination therapy in antiretroviral-naive adults with HIV: Planned interim 48 week results from SPRING-1, a dose-ranging, randomised, phase 2b trial. *Lancet, 12*(2), 111–118.
5. Cahn, P., Pozniak, A., Horacio, M., Andrey, S., Brites, C., Andrade-Villanueva, J. F., Richmond, G., Buendia, C. B., Fourie, J., Ramgopal, M., Hagins, D., Felizarta, F., Madruga, J., & Reuter. (2013). Dolutegravir versus raltegravir in antiretroviral-experienced, integrase-inhibitor-naive adults with HIV: Week 48 results from the randomised, double-blind, non-inferiority SAILING study. *Lancet, 382*(9893), 700–708.
6. Min, S., Sloan, L., DeJesus, E., Hawkins, T., McCurdy, L., Song, I., Stroder, R., Chen, S., Underwood, M., Fujiwara, T., Piscitelli, S., & Lalezari, J. (2011). Antiviral activity, safety, and pharmacokinetics/pharmacodynamics of dolutegravir as 10-day monotherapy in HIV-1-infected adults. *AIDS (London, England), 25*(14), 1737–1745.
7. Stephen, H., Steven, J. S., Mathieu, M., Albert, J.-C., Yves, P., Stephen, H. H., & Peter, C. (2011). Structural and functional analyses of the second-generation integrase strand transfer inhibitor dolutegravir (S/GSK1349572). *Molecular Pharmacology, 80*(4), 565–572.

8. Johannes, C., Christoph, L., & Jürgen, K. R. (2011). S/GSK1349572, a new integrase inhibitor for the treatment of HIV: Promises and challenges. *Expert Opinion on Investigational Drugs, 20*(4), 537–548.
9. Parienti, J.-J., Fournier, A. L., Cotte, L., Schneider, M.-P., Etienne, M., Unal, G., Perré, P., Dutheil, J.-J., Morilland-Lecoq, E., Chaillot, F., Bangsberg, D. R., Gagneux-Brunon, A., & Prazuck. (2021). Forgiveness of Dolutegravir-based triple therapy compared with older antiretroviral regimens: A prospective multicenter cohort of adherence patterns and HIV-RNA replication. *Open Forum Infectious Diseases, 8*(7), 1–8.
10. Rudy, L. M., Eric, M., & Nthembelihle, S. N. (2020). The prevalence of selected non-communicable disease risk factors among HIV patients on anti-retroviral therapy in Bush-buckridge sub-district, Mpumalanga province. *BMC Public Health, 20*(1), 247.
11. Achwoka, D., Waruru, A., Chen, T.-H., Masamaro, K., Ngugi, E., Kimani, M., Mukui, I., Oyugi, J. O., Mutave, R., Achia, T., Katana, A., Ng'ang'a, L., & De Cock, K. M. (2019). Noncommunicable disease burden among HIV patients in care: A national retrospective longitudinal analysis of HIV-treatment outcomes in Kenya, 2003–2013. *BMC Public Health, 19*(372), 1354–3784.
12. Jennifer, M.-G., Pascal, G., & Till, B. (2017). The causal impact of ART on NCDs: Leveraging quasi-experiments. *The Lancet, 5*(1), 14.
13. Laston, G., Indres, M., & Khumbulani, H. (2020). Effects of HIV and non-communicable disease comorbidity on healthcare costs and health experiences in people living with HIV in Zimbabwe. *Southern African Journal of HIV Medicine, 21*(1), 1102.
14. Getahun, Z., Azage, M., Abuhay, T., & Abebe, F. (2020). Comorbidity of HIV, hypertension, and diabetes and associated factors among people receiving antiretroviral therapy in Bahir Dar city, Ethiopia. *Journal of Comorbidity, 10*(2020), 1–12.
15. Mgaywa, G. M. D. M., Kazuhiko, M., Ehimario, U. I., Naoko, S. M., Michael, M., Julius, C. M., Dereje, H., Godfrey, M. R., & Masahiro, H. (2013). Non-communicable diseases in antiretroviral therapy recipients in Kagera Tanzania: A cross-sectional study. *Pan African Medical Journal, 16*(84), 2831.
16. Harald, W., Christos, N., Lia, B., & Alexander, B. L. (2022). Machine learning prediction of hypoglycemia and hyperglycemia from electronic health records. *JMIR Formative Research, 6*(7), e36176.
17. Lelani, C., Lisa, B., Jan-Walter, D. N., Till, B., Pascal, G., & Sebastian, V. (2019). HIV, antiretroviral therapy and non-communicable diseases in sub-Saharan Africa: Empirical evidence from 44 countries over the period 2000 to 2016. *Journal of the International AIDS Society, 22*(7), 1–8.
18. Sheila, K., Mwesigire, D., & Henry, M. (2019). Prevalence of non-communicable diseases among HIV positive patients on antiretroviral therapy at joint clinical research Centre, Lubowa, Uganda. *PLoS One, 14*(8), 1–11.
19. Milena, M., Sylvia, W., & Shannon, G. (2017). Dolutegravir-induced hyperglycaemia in a patient living with HIV. *Journal of Antimicrobial Chemotherapy, 73*(1), 258–260.
20. Henry, Z., Freddy, E. K., Helen, B. N., Nakitto-Kesi, D., Jacquellyn, N. S., & Ronald, K. (2021). Dolutegravir-based ART: Exploring patient safety and acceptability following National Roll-out in Uganda. *Research Square, 1*(2021), 1–22.
21. José, M. G., Lambert, A., Graeme, M., Laura, W., Margaret, J., Domingo, P., Julie, F., Esteban, M., Hans–Jürgen, S., Giovanni, G., Mar, M., Mark, G., Stephane, D. W., Eric, F., & Stefan, E. (2017). Switching from a ritonavir-boosted protease inhibitor to a dolutegravir-based regimen for maintenance of HIV viral suppression in patients with high cardiovascular risk. *AIDS, 31*(18), 2503–2514.
22. Jamison, N., Turner, M., Carmen, B., Peter, R., Bryan, S., Sally, B., Todd, H., Raffanti, S., Hass, D. W., Sterling, T. R., & Koethe, J. R. (2017). Brief report: Weight gain in persons with HIV switched from Efavirenz-based to integrase Strand transfer inhibitor-based regimens. *Journal of Acquired Immune Deficiency Syndromes, 76*(5), 527–531.

23. Menard, A., Meddeb, L., Tissot-Dupont, H., Ravaux, I., Dhiver, C., Mokhtari, S., Tomei, C., Brouqui, P., Colson, P., & Stein, A. (2017). Dolutegravir and weight gain: An unexpected bothering side effect? *Open Forum Infectious Diseases, 31*(10), 1499–1500.
24. Todd, H. (2018). Factors associated with insulin resistance in adults with HIV receiving contemporary antiretroviral therapy: A brief update. *Current HIV/AIDS Reports, 15*(3), 223–232.
25. Lucia, T., Paolo, B., Elena, R., Giancarlo, O., Nicola, S., Barbara, M., Giuseppe Vittorio, D. S., Giordano, M., Giovanni Francesco, P., Layla, P., Benedetto Maurizio, C., Leonardo, C., & Federi. (2020). Factors associated with weight gain in people treated with Dolutegravir. *Open Forum Infectious Diseases, 7*(6), 195.
26. Murray, M. M., & Harpe, S. E. (2020). 106. META-INSTI: Metabolic adverse events following integrase strand transfer inhibitor administration in spontaneous adverse event reports. *Open Forum Infectious Diseases, 7*(1), 182.
27. Sarah, M., & Eric, W. (2020). AIDS 2020 conference: Highlights and new data on HIV treatment. In *AIDS 2020*, San Francisco, California.
28. Lamorde, M., Atwiine, M., Owarwo, N. C., Ddungu, A., Laker, E. O., Mubiru, F., Kiragga, A., Lwanga, I. B., & Castelnuovo, B. (2020). Dolutegravir-associated hyperglycaemia in patients with HIV. *The Lancet HIV, 7*(7), 461–462.
29. Cogliandro, V., Lapadula, G., Squillace, N., Soria, A., Vitiello, P., Rossi, M., Colella, E., Columpsi, P., Migliorino, G. M., & Bonfanti, P.. (2020). New cases of diabetes dolutegravir-associated in HIV patients: A case series. In *ICAR2020*, Bologna.
30. Mwebesa, H., & Musinguzi, J. (2021). Ministry of Health. [Online]. Retrieved October 5, 2021, from https://differentiatedservicedelivery.org/Portals/0/adam/Content/HvpzRP5yUUSdpCe2m0KMdQ/File/Uganda_Consolidated%20HIV%20and%20AIDS%20Guidelines%202020%20June%2030th.pdf
31. Nestoras, M. N., Mohammed, A. S., Ahmed, S. F., Sam, S., Shamil, F., John, M., Mihail, Z., Suchi, S., & Sherita, G. H. (2021). Development and validation of a machine learning model to predict near-term risk of iatrogenic hypoglycemia in hospitalized patients. *JAMA Network Open, 4*(1), e2030913.
32. Ashenafi, Z. W., Eirik, Å., Ståle, W., David, A., Lena, M., Taxiarchis, B., & Gunnar, H. (2019). Data-driven modeling and prediction of blood glucose dynamics: Machine learning applications in type 1 diabetes. *Artificial Intelligence in Medicine, 98*(2019), 109–134.
33. World Health Organisation. (2013). *Consolidated guidelines on the use of antiretroviral drugs for treating and preventing HIV infection. Recommendations for a public health approach 2013*. World Health Organisation.
34. Melissa, A. J., Elizabeth, S. A., Constantine, F., & Philip, L. J. (2011). Multiple imputation with multivariate imputation by chained equation (MICE): What is it and how does it work? *International Journal of Methods in Psychiatric Research, 20*(1), 40–49.
35. Fernández, A., Garcia, S., Herrera, F., & Chawla, N. V. (2018). SMOTE for learning from imbalanced data: Progress and challenges, marking the 15-year anniversary. *Journal of Artificial Intelligence Research, 61*(1), 863–904.

Extraction of Numerical Facts from German Texts to Enrich Internal Audit Data

Gerrit Schumann and Jorge Marx Gómez

1 Introduction

One of the main tasks of internal auditing is to examine processes and structures throughout the company. In the course of these audits, auditors must gather information, reconcile it with other data points, and identify anomalies or discrepancies. In order to avoid sampling when dealing with ever-increasing amounts of data, internal auditing could nowadays already make use of technologies that enable the auditing of all available data instances (full-population testing) [1]. However, these technologies focus primarily on the company's structured data. Unstructured data, such as texts from business documents, on the other hand, are often still processed manually by auditors, which—due to the volume of data—ultimately leads to sampling again. For a company, sampling also always involves a "sampling risk," i.e., the risk that relevant information could be found in the transactions that were not selected for the audit [2]. Since text data constitutes a significant proportion of enterprise data, failure to include enough important information it contains represents a correspondingly high sampling risk.

To address this issue, natural language processing (NLP) techniques can be used to automatically extract relevant information contained in the text of business documents and transform it into a structured schema. In this way, the information can subsequently be processed by common audit tools, which not only saves the auditor a considerable amount of time but also allows all data to be taken into account.

A subcategory of NLP that can be used to accomplish such a task is "Open Information Extraction / OIE" [3]. OIE can be used to identify relationships in texts

G. Schumann (✉) · J. Marx Gómez
University of Oldenburg, Oldenburg, Germany
e-mail: gerrit.schumann@uol.de

J. Marx Gómez et al. (eds.), *Artificial Intelligence Tools and Applications in Embedded and Mobile Systems*, Progress in IS,
https://doi.org/10.1007/978-3-031-56576-2_16

without relying on manually set labels of that relationship. Instead, unlexicalized features, such as part-of-speech / POS tags (e.g., verb, noun, and adjective) or dependency tags (e.g., subject and predicate) are used to learn how relationships are expressed in written text in general—independent of specific domains [4]. In this way, from a sentence like *"Barack Obama was born in 1961 in Honolulu."* concrete facts can be extracted, typically represented as relational tuples (*"Barack Obama," "was born in," "1961"*), (*"Barack Obama," "was born in," "Honolulu"*).

First introduced by Banko et al. [3], OIE has been taken up and further developed by a large number of studies. These studies can be distinguished primarily by their respective algorithmic approaches and the language in which they are applied. One possible approach is to create hand-crafted rules that are used to form the patterns or tuples from the sentences. The input for such rules can be, for example, dependency tags [4–9], part-of-speech tags [10, 11] or the output of a constituent parser [12]. Another approach is to develop an OIE model using a concrete training base. For example, Banko et al. [3] used features such as POS tags and noun phrases to train a classifier that can be used to extract triple facts from a text corpus. Akbik and Broß [13] trained a pattern learner who used link grammar [14] to automatically learn 46 patterns from an annotated corpus. Also using a pattern learner in combination with dependency parse trees, Wu and Weld [15] determined whether the shortest path between two noun phrases expressed a semantic relation. More recent approaches used neural networks for their OIE systems [16–19].

Most of these studies applied their approaches to English texts. For Germany, only Falke [7], Bassa et al. [4], and Klose et al. [20] developed and tested corresponding methods. These approaches typically build their annotation schemes on the basis of the subject or the predicate in order to create triples or quadruples for all facts contained in a sentence. Such annotation schemes are useful in many situations, but may reach their limits when entities and relations are in focus that are not necessarily characterized by subject or predicate tags. Such a case can occur when we are interested in entities and relations mentioned in the context of numerical values. For example, in the German sentence *"Im letzten Quartal 2019 haben wir einen Nettogewinn von 12,8 Mio. Euro erzielt,"* the term *"wir"* would be the subject according to its dependency tag. In the annotation scheme of many existing approaches, the numeric value would then be placed within the object part along with the entity to which it actually refers—namely the "Nettogewinn" (*"wir," "haben erzielt," "einen Nettogewinn von 12,8 Mio. Euro"*). While such an annotation would be a correct subject-verb-object triple, it is not suitable for automated processing by audit tools because the value (*"12,8"*) is still "trapped" in the object phrase. Therefore, it would be better to have an annotation scheme where the numeric value is always isolated and can be clearly assigned to the entity it actually refers to. In such an annotation scheme, a "relation-class" must then describe how exactly the value and the entity are related. In the case of the above example sentence, the direct relation is *"von."* However, this term has no meaning without the rest of the relation, namely: *"haben wir erzielt"* or more precisely *"haben wir im letzten Quartal 2019 erzielt."* The formation of such relations is not always trivial, since the individual terms are located at different places in the sentence and must be logically assembled for a meaningful phrase. This is especially important when several numeric values are mentioned within one sentence. For example, in the sentence *"Durch den Geschäfts- ausbau über alle operativen Segmente hinweg erhöhte sich die Bilanzsumme zum Jah- resende 2019 um 15,4 Mrd. EUR auf 256,6 Mrd. EUR."* ideally (*"die Bilanzsumme zum Jahresende 2019", "erhöhte sich um", "15,4 Mrd. EUR"*) and (*"die Bilanzsumme zum Jahresende 2019," "erhöhte sich auf," "256,6 Mrd. EUR"*) should result as correct triples. In the case of this sentence, the two values refer to the same entity. The only and at the same time most important difference is that one value represents the final state of the entity and the other represents the change of the entity. This distinction must necessarily be reflected in the relation phrases ("*... um*" | "*... auf*").

To address the above-mentioned challenges in the best possible way, an OpenIE approach is needed that focuses on numerical facts. To the best of our knowledge,[1] only Saha and Pal [21] have developed such an approach. However, they only use English texts, which means that a direct application to German texts is only possible with the use of additional translation steps. Moreover, it cannot be guaranteed that the quality of the results will remain the same. The goal of this paper is therefore to develop an approach that is able to realize a correct extraction of all entities and relations that refer to numerical values in German sentences. In this way, we want to enable auditors to extract numerical facts from textual business documents and convert them into a structured schema so that they can be processed by conventional audit tools. For this purpose, we have taken up the basic idea of rule-based information extraction and extended it to fit our goal of numeric-focused IE (Sect. 2). We then evaluated our approach on a text corpus consisting of 4637 sentences from 50 German annual reports (Sect. 3) and drew a conclusion on the results (Sect. 4).

2 Information Extraction Focusing on Numerical Values

To address the challenges described in Sect. 1, we developed an IE approach for German texts that focuses on identifying entities and relations that refer to numerical values. This approach is based on a hand-crafted, rule-based logic that makes use of both dependency tree tags and part-of-speech tags of a sentence. In addition to focusing on numeric values, we also restrict this study to values associated with a currency (Euro and Dollar).

The order in which the following subsections are presented corresponds to the actual order within the extraction process. Accordingly, after the pre-processing has been completed, first the "Value" is determined, then the "Entity" is determined, and finally the "Relation" is determined. The procedure of our IE approach is illustrated in Fig. 1 with a concrete example, which is described in subsections 2.1, 2.2, 2.3 and 2.4.

[1] At the time of the conference in November 2022.

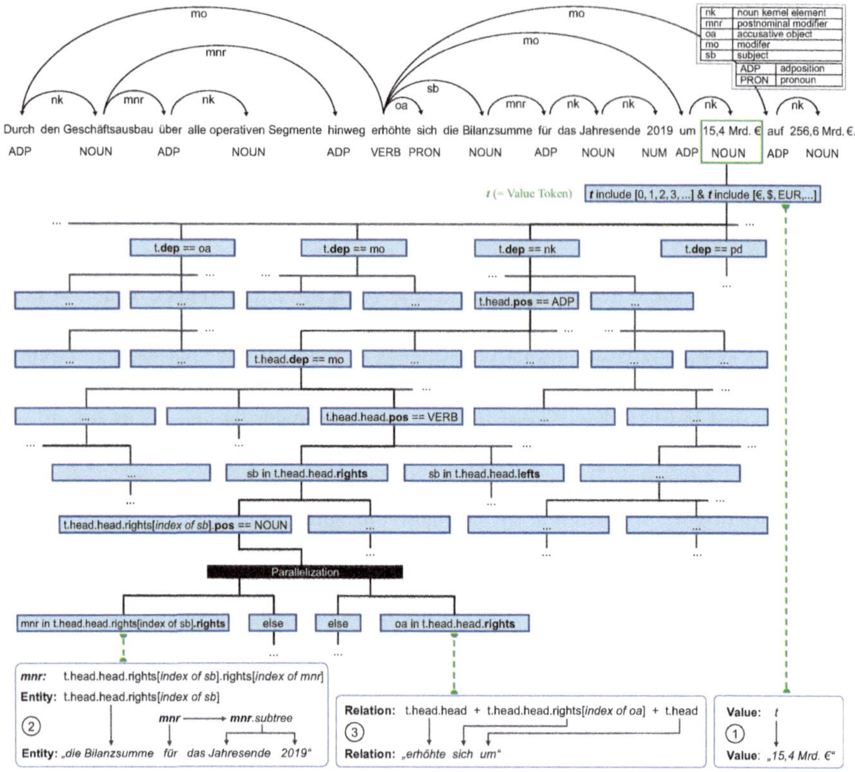

Fig. 1 Logic (–excerpt) for identifying the three target classes: Entity, Relation, and Value

2.1 Pre-Processing

As an initial step, the sentences contained in a given text corpus are segmented from each other and stored in a list. For this task, we used Spacys' Sentence Segmentation method.[2] From now on, the individual sentences are processed separately during a loop.

A first step in this processing is the transformation of the individual words and punctations into tokens, which then have POS- and dependency-tags. For this task, we used the German Spacy model "de_dep_news_trf,"[3] which is also based on the Transformer model "bert-base-german-cased."[4] Afterward, we combined—if possible—entities and noun chunks into composite tokens and made sure that

[2] https://spacy.io/usage/linguistic-features#sbd.

[3] https://spacy.io/models/de#de_dep_news_trf.

[4] https://huggingface.co/bert-base-german-cased.

there were no overlaps between them. In this way, the composite tokens *"den Geschäftsaufbau," "alle operativen Ge- schäftssegmente," "die Bilanzsumme," "das Jahresende," "15,4 Mrd. €,"* and *"256,6 Mrd. €"* would be created from the example sentence shown in Fig. 1.

In a final pre-processing step, we removed all bracket contents from a given sentence if numeric currency values are mentioned in it. We applied this filtering because numerical values in brackets are typically mentioned in a bullet-like manner and thus without a structure-giving sentence reference. The latter leads to the fact that the entity referring to this value often cannot be easily determined via the dependency tree tags.

2.2 Determination of the Numeric Value Token

For the determination of the numeric value token t only two conditions must be fulfilled. First, the token under consideration must be a noun or pronoun (t.pos == NOUN or t.pos == PRON) and second, it must contain a currency (€, Euro | $, Dollar) in addition to a numeric value (in our example: *"15,4 Mrd. €"*). The determination of the numeric value is the first step in our procedure, as the dependency tree is now traversed starting from this token (Fig. 1-①). This procedure is based on an example that can be found on the Spacy website about information extraction using dependency parsing.[5]

2.3 Determination of the Entity

The goal of this step is to determine the entity to which the numeric value actually refers. This entity may consist of a single noun or several compound noun chunks. To identify these tokens within the sentence, our approach first uses the dependency tag of the respective value token. Depending on the type of this tag, different paths must then be followed within the dependency tree (either upwards via "head"-element or downward into a subtree via "lefts"- or "rights"-element). In the case of the example sentence in Fig. 1, the numeric value (i.e., the starting token t) carries the dependency tag "nk," meaning that it is a noun kernel element of the associated head token. The rule defined for this case states that the information origin should be followed upwards, i.e., via head element (➔ t.head). From this level on, not only the dependency tag but also the POS tag of the respective token is considered. In our example sentence, the token at t.head is an "adposition" (ADP) and, according to its dependency tag, a "modifier" (mo). Also in this case, our rule

[5] https://spacy.io/usage/linguistic-features#:~:text=useful%20tool%20for-,information%20extraction,-%2C%20especially%20when%20combined.

system provides for following the information origin further up (→ t.head.head) and looking again at the dependency tag as well as the POS tag. However, since t.head.head is a verb according to the POS tag and thus a direct reference to a subject is likely, the dependency tag is considered secondary. At this point, the rule system provides for the first time that the search for the entity should be continued in a subtree. Therefore—if available—both the left and the right subtree of t.head.head are searched for a token that is a subject according to its dependency tag (sb). In our example sentence, such a subject token is located in the right subtree and is thus directly set as the entity if it is a noun (*"die Bilanzsumme"*). From now on, parallel to the identification of the relation class (Subsection 2.4), the system checks whether the entity is already complete or still needs to be extended with additional context. This is done by searching—depending on the availability—the right or left subtree of the previously determined entity token for a "postnominal modifier" (mnr). If it is found, it is added to the existing entity together with all tokens of the associated subtree (in our example sentence: [. . .] + *"für"* + *"das Jahresende"* + *"2019"*, Fig. 1-②).

2.4 Determination of the Relation

Once the final entity is determined, the next step is to find all relevant tokens to build the relation phrase. In the case of the example sentence shown in Fig. 1, the core element between the numeric value and the associated entity is the verb *"erhöhte."* In order to extend this first relation part meaningfully, the previously identified modifier *"um"* is now appended to the verb: *"erhöhte"* + *"um."* Depending on the type of verb and its context in the sentence, further additions may be necessary. For this reason, analogous to the procedure for determining the entity, the subtrees of the verb are also searched—this time, however, to find verb-extending tokens. In the case of our example sentence, such a verb extension is present as an accusative object (ao). To finally complete the relation phrase, the accusative object is placed between the verb and the modifier: *"erhöhte"* + *"sich"* + *"um"* (Fig. 1-③).

2.5 Scope of the Rule System

The example shown in Fig. 1 reflects a comparatively simple determination of entity and relation. In the case of long and nested sentences, our rule system follows longer paths through the dependency tree and thus becomes correspondingly more complex. Another factor that increased the scope and complexity of the rule system was our claim to build the information extraction as generically as possible. Accordingly, the size and complexity of the rule system also increased due to short or simple sentences, if their respective sentence structures were not considered before and thus had to be covered by new strands of logic. This process of generalizing the information extraction was done iteratively and resulted in a rule system that includes over 950 if-else statements.

3 Evaluation

3.1 Data Selection and Annotation

To evaluate our information extraction system, we needed a text corpus containing many numerical values that refer to specific entities within each sentence. Annual reports of listed companies meet this criterion and are also freely accessible. For this reason, we collected annual reports from a total of 50 German companies (from both the DAX and the MDAX) from 2021 and used them as a basis for our evaluation.

During pre-processing (Sect. 2.1), a total of 43,612 sentences were extracted from these 50 annual reports, of which 6937 sentences contained numeric values with currency signs. Since our approach is designed to determine the entities of at most two numeric values within a sentence, we also removed all sentences that contained more than two (currency-) values. In this filtering step, we also removed sentences where the value entities are pronouns that refer to previous sentences (e.g., *"**Diese** belaufen sich in Summe auf insgesamt rund EUR 23 Mio."* or *"Nach dem Dividendenbeschluss der Hauptversammlung am 3. Mai 2020 wurde **er** mit 29,40 € festgesetzt."*). After this final filtering step, 4637 sentences remained for our evaluation (Fig. 2).

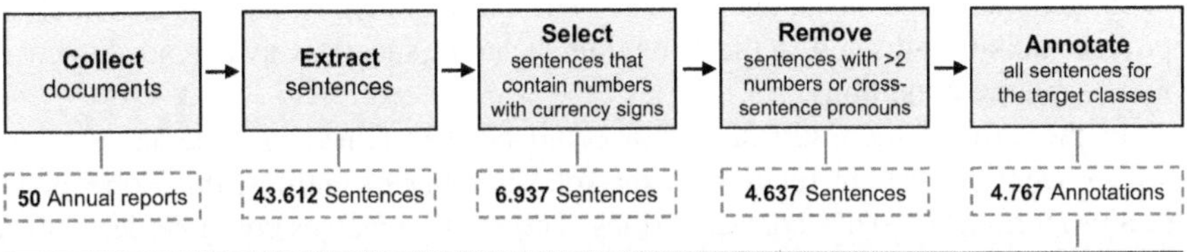

Sentence	Annotation		
	Entity	Relation	Value
Im Geschäftsjahr 2021 wurde ein Ergebnis vor Steuern in Höhe von 1,6 Mio. € erwirtschaftet.	Ergebnis vor Steuern	wurde im Geschäftsjahr 2021 erwirtschaftet, in Höhe von	1,6 Mio.
Gegenläufig reduzierten sich positive Marktwerte aus derivativen Sicherungsinstrumenten um -0,6 Mio. €.	positive Marktwerte aus derivativen Sicherungsinstrumenten	reduzierten sich um	-0,6 Mio.
Der Handelsumsatz sank im Vergleich zum Vorjahr auf 3,6 Mio. EUR.	Der Handelsumsatz	Sank auf	3,6 Mio.
Für 2023 wird ein Free Cashflow in Höhe von circa. 7,5 Mrd. € erwartet.	Free Cashflow	für 2023 erwartet, in Höhe von	7,5 Mrd.
Auf Grundlage des Aufhebungsvertrages würde eine Abfindungsleistung über 5,2 Mio. € vereinbart.	Abfindungsleistung	vereinbart, über	5,2 Mio.
Die zusatzversorgungspflichtigen Brutto-Löhne und Brutto-Gehälter betrugen im Berichtsjahr 19 Mio. €.	Die zusatzversorgungspflichtigen Brutto-Löhne u. -Gehälter	betrugen im Berichtsjahr	19 Mio.

Fig. 2 Procedure for the selection, filtering, and annotation of the evaluation data

The average length of these sentences is 24 tokens (words and punctuations), with a minimum of 6 tokens and a maximum of 95 tokens per sentence (median = 22 tokens). The total number of all tokens is 98,463, with 12,054 being unique. While the most frequently occurring POS tag is the pronoun (e.g., *welcher* and *deren*), the least frequently occurring POS tag is the subordinating conjunction (e.g., *weil*). The average dependency distance between each token and the root token of the sentence is 3.10.

For each of the 4637 sentences, we annotated the numeric value ("Value"), the entity to which the value actually refers ("Entity"), and the words by which the relation between Value and Entity is represented ("Relation"). Since 130 sentences contained two currency values, this step resulted in a total of 4767 annotations. Figure 2 also shows some short example sentences and their associated target classes. Afterward we let our script iterate over all sentences and compared the resulting target classes with our previously created annotations.

3.2 *Results*

Since the Value token within a sentence is identifiable by a simple logic and thus could be correctly determined in 100% of the cases, this section focused primarily on the evaluation of correct entities and relations. In this context, we also distinguish whether an entity or relation is correct or whether it was not determined at all. This is crucial because our information extraction approach follows handcrafted rules and determines an entity/relation only if the conditions within the sentence are covered by an appropriate logic. Although this may prevent some results from being generated, the advantage of such a strict rule system is that it is less likely to produce noisy results and, at the same time, it can be successively extended to cover more sentence conditions.

In the case of the entities, 82.0% could be determined in general, of which 98.2% were correct. In relation to the total number of entities, this corresponds to a proportion of 80.6% correct entities. The relation classes could be determined in 81.8% of all cases. Of these, 93.5% were correct, which corresponds to a ratio of 76.4% correct relations in total. Since our IE approach searches for the relation only after the entity has been found, the percentage of cases where both entity and relation were correctly identified is therefore also 76.4% (Table 1).

Table 1 Results of evaluations

	4637 Sentences	
	Entities	Relations
Entities and relations contained in all sentences	4767 (100%)	4767 (100%)
Determined by our IE-approach in general	3911 (82,0%)	3897 (81,8%)
Correctly determined by our IE-approach	3842 (80,6%)	3643 (76,4%)
Both correctly determined by our IE-approach	3643 (76,4%)	

Output of Information Extraction		
Entity	Relation	Value
Die Bilanzsumme für das Jahresende 2019	erhöhte sich um	15,4
Die Bilanzsumme für das Jahresende 2019	erhöhte sich auf	256,6
...	...	...

Transformation Variant 1	
Bilanzsumme_Jahresende 2019_erhöhte_um	Bilanzsumme_Jahresende 2019_erhöhte_auf
15,4	NaN
NaN	256,6
...	...

Transformation Variant 2			
Bilanzsumme_Jahresende_2019	erhöhte um	erhöhte auf	Value
1	1	0	15,4
1	0	1	256,6
...	...	...	...

Fig. 3 Possible data transformations to make the output processable for audit tools

3.3 Data Formats Suitable for Audit Tools

The final output resulting from our information extraction approach is in a three-column format (Fig. 3, left). Compared to the raw text data that was entered, this is already a structured data form that could in principle be further processed by audit tools. However, since only the Value column is in a numeric format, the remaining columns would have to be encoded into a format that can be interpreted by the respective tools. One option is to create a merged column identifier composed of the categorical columns "Entity" and "Relation." With this form of data transformation, no further encoding would be required since the cell contents are now exclusively numeric (Variant 1 | Fig. 3, top right). However, this transformation approach may result in a sparse data set in which many of the new column identifiers appearing only once or very rarely even though the single entity or relation already occurred during encoding.

One way to counteract this limitation would be simple one-hot-encoding, since recurring entities and relations are then considered separately and do not immediately result in a new column if they appear in different combinations within the IE output (Variant 2 | Fig. 3, bottom right).

4 Conclusion

In this paper, we have presented a rule-based approach to extract entities and relations, that refer to numerical values within a sentence. The distinctive feature of our approach is twofold. First, it specializes in the sentence structure of the German language. Second, it does not attempt to extract all types of facts within a sentence, but focuses exclusively on numerical values, from which it then determines the associated entities and relations. In this way, it is possible to determine the sometimes complex relationships through which entities and numeric values are connected in reality. The evaluation of our approach on a total of 4637 sentences, which were taken from 50 annual reports of German companies, shows that the entity could be determined in 80,6% of the cases and the relation in 76,4% of the cases.

For auditing practice, our approach can be helpful in that it favors an approximation to full-population testing by providing numerical facts from texts in an automated way. This, in turn, can reduce sampling risks as well as the overall cost of an audit. In addition to auditing, our approach can also be interesting for other domains where numerical values in texts need to be processed automatically. From a research perspective, our paper extends existing work from the field of information extraction with an approach that specializes in traversing even complex paths within the dependency tree to extract the entities and relationships that relate to numerical values in German sentences.

Despite the textual diversity enabled by 50 different documents and a total of 4637 individual sentences, a limitation of our approach is that we evaluated it on only one type of business document: Corporate annual reports. Moreover, our approach is still limited to numerical facts that refer to a currency (Euro / Dollar). Therefore, in order to enrich audit data in the best possible way, we plan both to extend the evaluation to other business documents and to extract entities and relationships of other types of numerical information such as quantities, sizes, weights, or distances.

References

1. Huang, F., No, W. G., Vasarhelyi, M. A., & Yan, Z. (2022). Audit data analytics, machine learning, and full population testing. *The Journal of Finance and Data Science, 8*, 138–144.
2. No, W. G., Lee, K., Huang, F., & Li, Q. (2019). Multidimensional audit data selection (MADS): A framework for using data analytics in audit data selection process. *Accounting Horizons, 33*(3), 127–140.
3. Banko, M., Cafarella, M. J., Soderland, S., Broadhead, M., & Etzioni, O. (2007). Open information extraction from the web. In *Proceedings of the International Joint Conference on Artificial Intelligence, IJCAI '07* (pp. 2670–26760). AAAI.
4. Bassa, A., Kröll, M., & Kern, R. (2018). GerIE—An open information extraction system for the German language. *Journal of Universal Computer Science, 24*(1), 2–24.
5. Akbik, A., & Löser, A. (2012). Kraken: Nary facts in open information extraction. In *Proceedings of the Joint Workshop on Automatic Knowledge Base Construction and Web-scale Knowledge Extraction, AKBC-WEKEX'12* (pp. 52–56). Association for Computational Linguistics.
6. Del Corro, L., & Gemulla, R. (2013). Claus IE: Clause-based open information extraction. In *Proceedings of the 22nd International World Wide Web Conference. International World Wide Web Conferences Steering Committee* (pp. 355–365). ACM.
7. Falke, T. (2016). Porting an open information extraction system from English to German. In *Proceedings of the 2016 conference on empirical methods in natural language processing (EMNLP-16)* (pp. 892–898). Association for Computational Linguistics.
8. Gamallo, P., Garcia, M., & Fernandez-Lanza, S. (2012). Dependency-based open information extraction. In *Proceedings of the 13th Conf. of the European Chapter of the Association for Computational Linguistics* (pp. 10–18). Association for Comput. Linguistics.
9. Gamallo, P., & Garcia, M. (2015). Multilingual open information extraction. In F. Pereira & Machado, P., Costa, E., Cardoso, A. (Eds.), *Progress in Artificial Intelligence. EPIA 2015. Lecture notes in Computer Science* (Vol. 9273, pp. 711–722). Springer.

10. Xavier, C. C., & de Lima, V. L. S. (2014). Boosting open information extraction with noun-based relations. In *Proceedings of the 9th International Conference on Language Resources and Evaluation, LREC'14* (pp. 96–100). ELRA.
11. Zhila, A., & Gelbukh, A. (2016). Open information extraction from real internet texts in Spanish using constraints over part-of-speech sequences: Problems of the method, their causes, and ways for improvement. *Revista signos, 49*(90), 119–142.
12. Bast, H., & Haussmann, E. (2013). Open information extraction via contextual sentence decomposition. In *2013 IEEE Seventh International Conference on Semantic Computing* (pp. 154–159). IEEE.
13. Akbik, A., & Broß, J. (2009). Wanderlust: Extracting semantic relations from natural language text using dependency grammar patterns. In *Proceedings of the 2009 semantic search workshop at the 18th Internet. World wide web conference* (pp. 6–15). ACM.
14. Sleator, D. D., & Temperley, D. (1995). Parsing English with a link grammar. arXiv preprint cmplg/9508004.
15. Wu, F., & Weld, D. S. (2010). Open information extraction using Wikipedia. In *Proceedings of the 48th Annual Meeting of the Association for Computational Linguistics, ACL'10* (pp. 118–127). Association for Computational Linguistics.
16. Cui, L., Wei, F., & Zhou, M. (2018). Neural open information extraction, arXiv preprint, arXiv: 1805.04270.
17. Roy, A., Park, Y., Lee, T., & Pan, S. (2019). Supervising unsupervised open information extraction models. In *Proceedings of the 2019 Conference on Empirical Methods in Natural Language Processing and the 9th International Joint Conference on Natural Language Processing* (pp. 728–737). Association for Computational Linguistics.
18. Stanovsky, G., Michael, J., Zettlemoyer, L., & Dagan, I. (2018). Supervised open information extraction. In *Proceedings of the 2018 Conference of the North American Chapter of the Association for Computational Linguistics: Human Language Technologies, Volume 1 (Long Papers)* (pp. 885–895). Association for Computational Linguistics.
19. Zhan, J., & Zhao, H. (2020). Span model for open information extraction on accurate corpus. In *Proceedings of the AAAI Conference on Artificial Intelligence (Vol. 34, No. 5)* (pp. 9523–9530). AAAI.
20. Klose C., Gui, Z., & Harth, A. (2022). Open information extraction on German Wikipedia texts. In Text2KG 2022: International workshop on knowledge graph generation from text, co-located with the ESWC 2022, May 05-30-2022, Crete, Hersonissos, Greece.
21. Saha, S., & Pal, H. (2017). Bootstrapping for numerical open IE. In *Proceedings of the 55th Annual Meeting of the Association for Computational Linguistics, Volume 2: Short Papers* (pp. 317–323). Association for Computational Linguistics.

From Process Mining to Enterprise Mining

Hans-Jürgen Scheruhn, Christian Reiter, and Elnur Bayramli

1 Introduction

In accordance with the Design Science research paradigm [1], this thesis examines the following research question: How can an Enterprise Resource Planning (ERP) system be analyzed so that the results of the analysis are sufficient for an almost all-encompassing view of a company in the context of digitization projects and so that the same results also accelerate the passage through phases 1 and 5 of the Process Life Cycle? [2].

To answer the research question, a term derived from data and process mining called enterprise mining is introduced and based on a literature review, technical data collection methods for application to ERP systems (using SAP as an example) are presented and tested with respect to goal achievement. The measurement of goal achievement is based on problem-oriented requirements, which are established in the context of the research approach, as well as on the enterprise architecture model Enterprise Online Guide, which is designed to illustrate both business and technical aspects of a company, such as business model, organizational structure, functions, processes, applications, and data, but also technologies and networks.

H.-J. Scheruhn · E. Bayramli (✉)
Hochschule Harz, Wernigerode, Germany
e-mail: hscheruhn@hs-harz.de; ebayramli@hs-harz.de

C. Reiter
Hamburg, Germany

J. Marx Gómez et al. (eds.), *Artificial Intelligence Tools and Applications in Embedded and Mobile Systems*, Progress in IS,
https://doi.org/10.1007/978-3-031-56576-2_17

2 Background

2.1 Process Mining

In the context of the Process Life Cycle, process mining is known to serve several purposes. In this context, Accrosi et al. [3] distinguish between the three types of process mining (detection (A), conformance checking (B), and extension (C)). All three types of process mining use event logs and can provide process-oriented insights on the following four aspects of analysis [cf. van der Aalst, 2016: 34]: Control flow, case, time, and organization.

Van der Aalst et al. [4] rate event logs according to five maturity levels (maturity level five has the best quality) and note that an event log for process mining must be at least maturity level three. Event logs used in this work, which originate from an ERP system, typically have maturity level three. Both technical and business knowledge are necessary in companies for implementation, whereby this turned out to be the greatest hurdle regarding process mining in the survey by Harmon and Garcia [5].

2.2 Enterprise Online Guide

The Enterprise Online Guide (also: EOG) according to Scheruhn et al. [2] is an information architecture and helps to structure and model differently categorized and classified information about a company to obtain an overall view of it. This is done using eight aspects of a company (also known as maps), each with four abstraction levels (also known as zoom factors). The abstraction levels extend the existing object-oriented and service-oriented architectures to include a consistent process orientation.

The EOG framework is characterized by a tabular representation with eight maps arranged horizontally and four zoom factors arranged vertically.

The four zoom factors symbolize the level of detail or depth of the process information. The first zoom factor symbolizes a high-level perspective on the information by providing an overview of the largest components. As the zoom factor is increased, an increasing low-level perspective is achieved. Accordingly, zoom factor 4 mainly describes the details of individual process components [6]. Following Scheruhn et al. [2], the following guiding questions can be derived per EOG map:

- Motivation (M): Which are the (internal) customers of my organization and which goals as well as strategies arise from a common motivation to improve the company processes? How does the corresponding Customer Journey look like?
- Organization (O): How is the organization structured (process-oriented)?

- Function (F): Which functions can the organization provide for the customers (process-oriented)?
- Process (P): What processes exist in the company? Which processes best align the organization's functions with the customer's needs and goals?
- Application (A): Which IT application is used to implement the organization's processes? (e.g., SAP, MS Access, and MS Excel).
- Data (D): Which data represent the processes and how are they structured (horizontally and vertically)?
- Technology (T): Which technological and social platforms are used (process-oriented) and how are they set up (customizing)?
- Network (N): How are these interconnected in a process-oriented way (technically and socially)?

The eight EOG maps mentioned in the list are combined into three superordinate EOG maps (Sustainability Map, Intelligence Map, and Ecosystem Map). The Sustainability Maps symbolize a sustainably operated business, e.g., based on a sustainability-balanced scorecard in the Motivation Map. The Digital Business Model Diagram describes nine different potentials (according to Osterwalder) of a sustainable enterprise digitization. The Intelligence Maps describe how humans act intelligently in their interaction with information systems. The Ecosystem Maps express how people in the company are networked with each other by means of technologies [7].

3 Design of the Solution Environment

3.1 Enterprise Mining as an Extension of Process Mining

The term "enterprise mining" in the context of process mining or as an extension of it is not widely used worldwide and, from the authors' point of view, offers room for research activities. One possible reason for this is that the discipline of process mining only emerged at the end of the 1990s and is thus comparatively young. Most user companies are still at the beginning of their process mining projects and the majority focus on the first type of process mining (detection) [8]. Nevertheless, using the same technology as for process mining, the narrative of an enterprise mining can be aimed at as an extension of process mining, which is not only primarily about collecting information about the process view, but those about all aspects of a company that are relevant in the context of digitalization or enterprise (re-) engineering. The goal of enterprise mining is to create a digital twin of a company to use it, for example, to find optimization potential or to holistically plan the ongoing digitization of the company.

In this context, Hess [9] describes a total of six aspects (business ecosystem, process and task mining, organizational mining, customer journeys, material and goods flows, data, and information flow) that could be considered in the context of

		SUSTAINABILITY MAP				INTELLIGENCE MAP		ECOSYSTEM MAP	
		Motivation	Organization	Function	Process	Application	Data	Technology	Network
Corporate Management	1	Company goals Business model	Society	Products and services Customer Journey	Scenarios	Value-added areas	Big Data	Server landscape	IT service model Network infrastructure
Division Management	2	Departmental goals	Departments		Processes and variants	Department functions	OLAP vs. OLTP	Software landscape	
Area Management	3	Workplace targets	User, Sys. org, Profiles, rollers		Process steps, tasks	Event, P.-step, Data, IT, Roles	Master & transaction data	System components, Customizing, Programs	SOD after authorization
Compliance	4	Target agreements	Employment contracts, User Performance	Service Level Agreements, Performance	KPI (e.g. lead time)	Mask structure	Log data, Table usage	Licenses, Usage protocols	SOD After use

Fig. 1 Enterprise mining in the context of Enterprise Online Guide [own representation based on [2, 10]]

enterprise mining. However, from the authors' point of view, what is missing in the existing literature is a fully described concept for an enterprise mining. Therefore, in the next chapter, the Enterprise Online Guide is proposed as a framework concept for enterprise mining.

3.2 EOG Framework as a Framework Concept for Enterprise Mining

Fig. 1 describes a proposal of the authors for enterprise mining based on the EOG and shows which information should optimally be collected within its eight maps and four zoom factors in the context of an actual situation analysis. The requirements defined in the EOG take into account the specifications in the convention manual [10] and are also compatible with those of Hess [9]. The same applies to the analysis aspects of process mining defined by van der Aalst: (1). control flow (EOG: Process Map), (2). case (EOG: Function Map), (3). time (EOG: Application Map), (4). organization (EOG: Organization Map) [11]. In the individual cells of the EOG, a decomposition of the previously presented guiding questions per EOG map into the individual EOG zoom factors can be seen.

4 Demonstration

In this chapter, three ways of technically implementing enterprise mining based on existing tools (msgFIT, SAP Signavio Process Intelligence, SAP Signavio Process Insights) are presented. All of them have in common that they are specifically designed for the collection of information from SAP ERP systems and read out specific database tables or event logs. For each of the options shown, an EOG coverage rate is also determined, which provides information about the ratio between the determined and non-determined EOG cells.

4.1 *Redocumentation with msgFIT (Process Mining Types B and C)*

From a theoretical point of view, software for redocumentation can be divided into two components: (1.) Reverse Engineering and (2.) Reverse Business Engineering.

The reverse engineering component analyzes the technical aspects regarding structures, states, and behaviors in the ERP system. This means, for example, the collection of information regarding the transactions used and the customizing settings from a client [12].

Redocumentation is assigned to process mining types B and C because an already existing process library is compared with the information about the customized settings and the actual transaction usage from the ERP system and as a result a process library (also: process documentation) adapted to the analyzed company is generated with the system-supported processes.

In this work, the tool-supported redocumentation service "msgFIT" of the IT service provider msg services ag from Munich was used for demonstration purposes due to its long deployment experience and specialization in SAP ERP systems [13].

From the msgFIT usage analysis data in Excel format, the results of the above analysis aspects could be mapped to the individual cells of the EOG framework in the Fig. 2 to estimate the scope of data collection using msgFIT (redocumentation) for enterprise mining and to determine an EOG coverage rate.

This shows that most of the information from msgFIT can be assigned to the Intelligence and Ecosystem Map of the EOG and that the IT department of a company is accordingly suitable as the primary target group for the msgFIT results. In this context, the EOG coverage rate (rounded percentage of filled EOG cells) is 59%.

		SUSTAINABILITY MAP				INTELLIGENCE MAP		ECOSYSTEM MAP	
		Motivation	Organization	Function	Process	Application	Data	Technology	Network
Corporate Management	1		Society		Scenarios	Value-added areas			
Division Management	2		Departments		Processes and variants	Department functions	OLAP vs. OLTP	software landscape, Proprietary developments	
Area Management	3		User, Sys.Org, People, profiles/roles		Process steps, tasks Documents	Tcode Sys-Org. User	Trunk & Movement D.,	System components, customizing	SOD after authorization
Compliance	4		User Performance			Tcode KPIs such as dialog steps	Table usage	Licenses, Usage protocols	SOD After use

Fig. 2 EOG coverage of msgFIT (redocumentation) [own illustration]

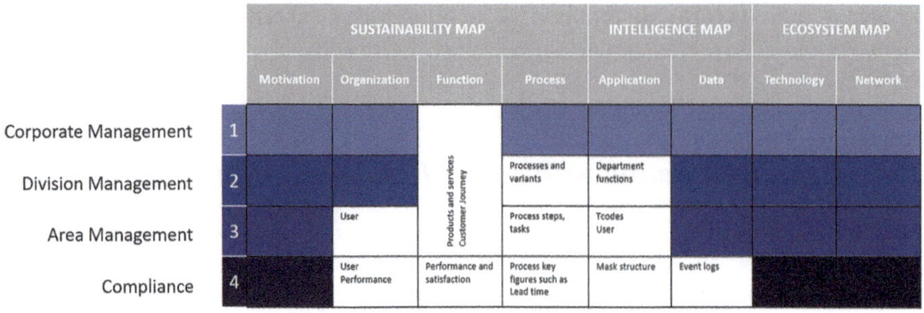

Fig. 3 EOG coverage of process mining [own illustration]

4.2 *Process Mining with Signavio*

To determine the EOG coverage rate and thus the scope of data collection for enterprise mining, the information collected using process mining was mapped into the individual cells of the EOG framework. The result can be seen in the Fig. 3.

In contrast to redocumentation, process mining reveals an information overhang in business aspects of the EOG (sustainability map). Accordingly, the business departments of a company are suitable as the primary target group for the results from classic process mining because they are typically interested particularly in the detailed information about process, product, and service performance. The EOG coverage rate for process mining (rounded-up percentage of filled EOG cells) is 41%. When comparing the EOG coverage rate of the redocumentation, it is important to note that the coverage rate of process mining can be extended, for example, in the Data, Technology, and Network Map of the EOG by means of additional programming or customizing in the process mining tool. However, since this is associated with additional (customizing) effort from the authors' point of view [2], the focus in the implementation of process mining was on those cells of the EOG that are supplementary to the redocumentation on the one hand but can also be implemented using the standard functionalities of the process mining tools (without additional programming) on the other hand.

4.3 *Supplementation by Means of SAP Process Insights*

SAP Process Insights is offered by SAP SE as a cloud service and has the advantage that user companies (like redocumentation) do not have to plan any customizing efforts for implementation because the analysis tool is specially adapted to SAP ERP or SAP S/4HANA and collects predefined key figs [14]. Based on a demo version, the information generated using SAP Process Insights was assigned to the EOG maps and the result is documented in Fig. 4. In this figure, an information

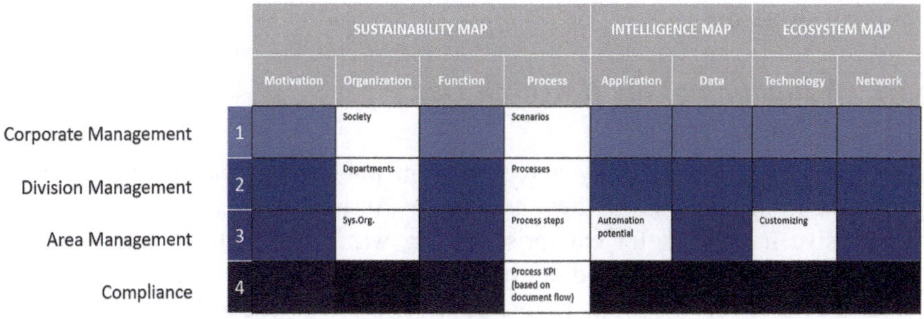

Fig. 4 EOG coverage of SAP process insights [own illustration]

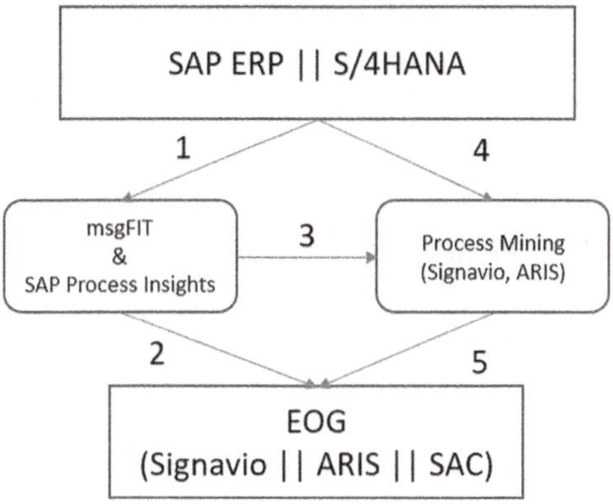

Fig. 5 Enterprise mining process model [own illustration based on [2]]

overhang can be seen in the business maps of the EOG, whereby the EOG coverage rate in this case is (rounded down) 28%.

4.4 Procedure for Enterprise Mining

The authors propose the enterprise mining approach shown in Fig. 5.

Initially, the information is collected using msgFIT and SAP Process Insights and the information is transferred to the EOG using SAP Solution Manager and predefined interfaces of modeling tools (Signavio, ARIS) (Step 2).

The EOG is then covered by 59% and serves as a central storage location for all information. In the third step, process mining (types B and C) is set up based on the information already available about customizing and process variants and then applied to the ERP system (Step 4). The results are transferred to the EOG in the fifth step, which checks the information generated in the first step for conformance and expands it. In this state, the EOG is 75% covered.

5 Evaluation

5.1 Validation of the Research Gap

Unlike process mining, only a few relevant publications on enterprise mining emerged from the authors' literature search. As indicated earlier, no fully comprehensive framework for enterprise mining was found in this context. Similar approaches have been pointed out by other authors (van der Aalst and Hess), but with the result of this work further analysis aspects based on EOG have been added. Accordingly, since this is the first publication that introduces the EOG as a framework for an enterprise mining, considers all mentioned maps of the EOG and at the same time also analyzes tools for the implementation, the novelty is given. In addition to the novelty, the necessity was previously justified, among other things, with optimization potentials in implementation projects of ERP software.

5.2 Design of the Solution Environment

The validation of the design of the solution environment is based on the problem-oriented requirements A.1 and A.2 (Table 1).

To provide a holistic overview of an enterprise, the EOG with eight maps and four zoom factors each was introduced as a framework for enterprise mining and used as a basis for design (A.1). As shown in Fig. 6, the validity with regard to the requirement of the EOG maps results, for example, from a comparison with other common enterprise and information architectures (like Zachman, TOGAF, ARIS, ArchiMate 3.1, RAMI 4.0, SCOR, and APQC), which show how individual components of other architectures can be recognized in the EOG [15].

To address different target groups in the context of an ERP project (A.2), the EOG Maps also offer the possibility to address three target groups: those with a focus on business aspects (Sustainability Maps), points of contact between business and technology (Intelligence Maps), and with a focus on technology (Technology Maps).

Table 1 Requirements for the solution environment [own representation]

ID	Description
A.1	The selected aspects of analysis are intended to provide a holistic overview of the company
A.2	Addressing different target groups in an ERP project with both business and IT metrics

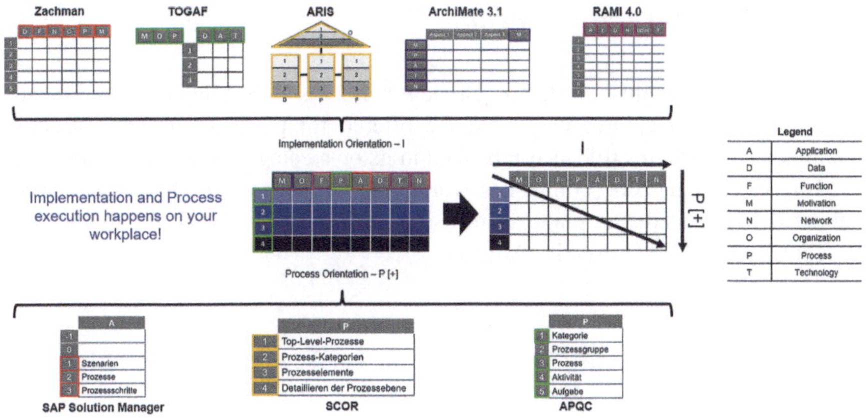

Fig. 6 Comparing EOG with other information architectures [15]

5.3 Prototypical Instantiation

To demonstrate the technical feasibility, the instantiation of the enterprise mining solution based on the EOG and the technical tools for data collection msgFIT, Signavio Process Intelligence as well as SAP Process Insights was tested within the scope of this work. For this purpose, an efficiency-oriented alignment of the tools was explained from the point of view of the authors, which is also intended to overcome the hurdles of process mining. One advantage of the chosen tools is that they each have their own thematic focus (be it business or technical) and complement each other with different metrics. An EOG coverage rate of 75% was achieved and at the same time a starting point for further research activities to increase the coverage rate was created.

The coverage rate indicates how many cells of the EOG collected information could be assigned. Another dimension for assessing the coverage rate, which was not initially considered in this work, is the number of data objects collected per EOG cell. However, this requires a well-founded definition of the individual EOG cell contents including the individual information objects, which do not yet exist in this form and may be a result of future research activities.

A critical aspect regarding tool selection is that msgFIT and SAP Process Insights can only be used on ERP systems from the manufacturer SAP because their data extractors—in contrast to the process mining tools—are geared to SAP's software architecture. In addition, only digitized processes are taken into account during data collection, which only helps companies with a low digitization rate to a limited extent. To counteract this, the approach of Berghaus et al. [16] can be integrated into the overall concept developed here, for example, as part of a further iteration of the DSR process, to be able to identify digitization or optimization potential of non-digital business processes.

5.4 Use in Real Environments

The tools used in this work are already being used beneficially (but mostly separately) in real environments and are offered for purchase by the respective manufacturers. The msgFIT is established in its own category and has already been used by more than 2000 companies worldwide [13]. The EOG has already been validated in an academic context and is offered as a framework for global SAP teaching by the SAP University Alliance.

6 Conclusion

The application of the presented concept enables on the one hand a holistic overview of a company and on the other hand serves as a basis for data-driven decisions and thereby optimizes phases 1 and 5 of the process cycle.

Using eight maps and four zoom factors of the EOG, a scientifically sound overall view of a company is given. Appropriate data collection methods were consolidated and a 75% coverage rate of the EOG was achieved. For better applicability in real environments, an enterprise mining approach was proposed. The result basically aims at a holistic overview of an enterprise. However, a complete coverage of the EOG with the available methods was not possible at first. Therefore, it suggests itself to identify possibilities for the so far non-collected EOG cells in future research and to create an own technical solution if necessary. A further result can be the extension of the concept based on the consideration of a further dimension for the determination of the EOG coverage rate, whereby for this first the accompanying conditions must be created regarding the EOG definition. Furthermore, it should be examined whether this framework is suitable for implementation in the new SAP Process Explorer in connection with the SAP One Process Acceleration Layer (OPAL) with a possible extension by "Sustainability" or "Intelligence" maps from EOG.

References

1. Hevner, A., March, S., Park, J., & Ram, S. (2004). Design science in information systems research. *MIS Quarterly, 28*, 75–105.
2. Scheruhn, H.-J. (2021). Retrieved February 05, 2022, from using enterprise maps to_finally_find_its_way_better_in_ERP_systems: https://www.youtube.com/watch?v=WouErEEiD1I&ab_channel=MichaelGr%C3%B6schel
3. Accorsi, R., Ullrich, M., & der Aalst, W. M. (2012). *Informatics_spectrum_35_5_2012: Process mining*. Springer-Verlag. https://doi.org/10.1007/s00287-012-0641-4
4. Aalst, W. V., & IEEE, P. T. (2011). Process mining manifesto. In *Business process management workshops 2011, lecture notes in business information processing* (Vol. 99). Springer.
5. Harmon, P., & Garcia, J. (2020). The state of business process management.

6. Scheruhn, H.-J. (2012, November 19). SAP community blog. Retrieved December 28, 2021, from https://blogs.sap.com/2012/11/19/online-process-management-at-gbi-part-2/
7. Scheruhn, H.-J., Weidner, S., & Müller, S. (2021a). Case study: Enterprise online guide.
8. Gartner Inc. (2021). Market guide for process mining. https://www.gartner.com/doc/reprints?id=1-2816N9FO&ct=211112&st=sb
9. Hess, H. (2021, September 27). Software AG Blog: Enterprise mining after the tipping point. Retrieved January 07, 2022, from https://blog.softwareag.com/enterprise-mining
10. Thielert, B. (2011). ARIS conventions.
11. Aalst, W. V. (2016). *Process mining, data science in action*. Springer.
12. Reiter, C. (2002). Model-based SAP R/3 Redocumentation. In *ARIS in practice: design, implementation and optimization of business processes* (1st ed., pp. 219–240). Springer Berlin Heidelberg.
13. msg. (2021). msgFIT: Our analysis for your successful growth. Retrieved December 31, 2021, from https://www.msg-treorbis.de/unsere-loesungen/msgfit-s-4hana-conversion
14. SAP. (2022a). Retrieved January 07, 2022, from https://www.sap.com/germany/products/process-insights.html
15. Graf von Hardenberg, M. (2022). SAP academic community conference: Comparing Enterprise architectures.
16. Berghaus, G., Kessler, R., Dmitriyev, V., & Gomez, J. M. (2018). *Identifying the digitization potential of non-digital business processes*. Springer Fachmedien Wiesbaden GmbH, part of Springer Nature.

CNN with Attention-Guided Concatenation for Improved Image Restoration

Gladys Mange, Ronald Waweru, Michael Kimwele, and Jorge Marx Gómez

1 Introduction

Image restoration is a type of image processing which seeks to modify degraded images to clean images. Zha et al. [1] reiterate that image restoration is a fundamental problem in image processing. Image restoration aims to reconstruct the original image x from its degraded observation y as precisely as possible. The degradation model for image restoration can mathematically be modeled as:

$$y = Hx + e \tag{1}$$

where e is the additive white Gaussian noise vector and H is a degradation matrix.

Image restoration, according to Sun S. [2], is the processing of images to recover the original image's content from a degraded image. Degradation occurs in image such as distortion and blur, Maru, M., & Parikh, M.C [3]. Image restoration algorithms are essential in criminal investigation, remote sensing surveys, military science, biological science, medical imaging, and traffic monitoring, Farhan Rabab [4]., and Khare Charu, [5]. In modern video surveillance systems, image restoration is crucial for criminal inquiry, Sun S [2]. However, due to the motion and low resolution of the image, image degradation is inevitably generated during data acquisition and transmission, which affects the quality of the original image. As a result, image restoration is essential in addressing the issue of image degradation. Image restoration is different from image enhancement in that the former involves

G. Mange (✉) · R. Waweru · M. Kimwele
Jomo Kenyatta University of Agriculture and Technology, Nairobi, Kenya

J. Marx Gómez
University of Oldenburg, Oldenburg, Germany

J. Marx Gómez et al. (eds.), *Artificial Intelligence Tools and Applications in Embedded and Mobile Systems*, Progress in IS,
https://doi.org/10.1007/978-3-031-56576-2_18

transforming distorted images into high-quality ones, whereas the latter seeks to make the image beautiful and pleasing to the eyes, Mohammed O. et al. [6]. Tian C [7]. found that many researchers have proposed combining image features from different CNN architectures, and most of these methods adopt concatenation as the mechanism to combine these features. However, simply concatenating the outputs from the different CNNs is unsatisfactory.

This research seeks to develop an optimized CNN model by combining two networks to increase the width hence features that can be extracted from the input images and therefore improve the performance of the combined network. This approach uses two attention Modules at the end of each CNN to select the features that impact the module's performance. The selected features are then concatenated to produce the output image. The outcome of this research is an improved CNN model that introduces the Attention module before concatenation operation to improve the feature selection process. It combines two networks to increase the width and, therefore, the receptive field (therefore increasing the learning ability of the network), validation is done on the two models: (CNOATT) and (CWATT) using CBSD68 and Kodak24 datasets and the latter performs better in both datasets. This is a clear indication that introducing an Attention module before concatenation of final output features yields better results.

2 Literature Review

A larger receptive field implies better denoising performance and efficiency. This has motivated many researchers to promote CNN denoising models by increasing the receptive field. The enhancement of CNN by use of dilated convolutions and a feed-forward mechanism was suggested by Chen C. et al. [8] in order to permit more beneficial features for noise filtering. By locating the closest point on the GAN manifold and retrieving latent vectors by condensing distances in image space, Tripathi S [9]. offered to denoise corrupted images; nevertheless, Z. Yue et al. [10] pointed out that the instability of training the GAN is a disadvantage. Z. Yue et al. [10]. Tai, Y. et al. [11] developed a profound, persistent memory network (MemNet) that explicitly introduces a memory block composed of a recursive unit and a gate unit to mine persistent memory through an adaptive learning process, but the memory block which is the main building block of MemNet does not fully utilize the features from the convolutional layer, Y. Zhang et al. [12], Mao X. et Al. [13] used the skip connections between convolutions and deconvolutions to eliminate image degradation. Liu P [14]. fused the wavelet transform and a U-Net to extract detailed information about the corrupted image but nevertheless, the network needs a major modification to maintain proper integrity, Bhavani M.D. et al. [15]. Zhang K. et al. [16] proposed a fast and flexible denoising convolutional network (FFDNet) which works on downsampled sub-images, achieving a good trade-off between inference speed and denoising performance but FFDNet has high time complexity, M. Yadav and N. Tiwari [17].

2.1 Feature Fusion

Feature fusion combines different layers or branches and has been studied widely, for example, in inception networks, Szegedy C et al. [18, 19] and Srivastava, R. K. et al. [20] fuse the outputs of the filters with multiple sizes of the same level. Residual networks (Res-Net) He, K. et.al [21]., use the identity mapping features and residual learning features at the output via short skip connections enabling the training of very deep networks in feature pyramid networks (FPN), however, its structure lead to buildup of mix of feature representation at each layer despite the fact that some features learned in earlier layers may no longer provide useful information in later layers, Targ S. et al. [22] Lin, T. Y. et al. [23], Ronneberger O. et al. [24] fuse low-level features and high-level features via long skip connections to obtain high resolution and semantically robust features, however, this method suffers feature-level imbalance among multi-level features. Since the detection head predicts scores, offsets independently on each feature, it leads to disparity between the head branches, Vo, X. T.et al. [25].

The above deep learning methods have two shortcomings (1) They do not make full use of the influence of an enlarged receptive field and (2) Do not take into account selection of features that are of importance in the denoising process. This research seeks to address these shortcomings in the proposed model. Feature fusion has not been appropriately addressed in the past and has been implemented in simple operations such as additions or concatenation, which might not be the best choice.

2.2 Dilated Convolutions

Traditional CNNs used pooling operations to reduce the dimensionality of original images. However, the results suffered from information loss. Enlarging the receptive field was a practical approach to addressing this issue. Increasing depth and filter size are prevalent ways to enlarge the receptive field. However, increasing depth may lead to network performance degradation. Enlarging the filter size will increase the number of parameters and computational cost.

2.3 Attention Mechanism

The extraction and choice of suitable features are paramount in the promotion of image processing applications, Li, X. et al. [26], but challenging at the same time. Attention mechanism has been proposed recently due to the capability to highlight important features by scoring each dimensional feature of the inputs and has been used widely in image restoration.

Anwar, S. et al. [27] formulated attention as follows: Let fc be the output features of the last convolutional layer having c feature maps of size h × w; global average pooling will reduce the size from h × w × c to 1 × 1 × c as:

$$gp = \frac{1}{hxw} \sum\nolimits_{i=j}^{h} \sum\nolimits_{i=1}^{w} fc\,(i, j) \tag{2}$$

where Fc(i,j) is the feature value at position (i,j) in the feature maps.

3 Methodology

The proposed CNN model (Fig. 1) is developed by combining two networks to increase the width and hence improve the combined network's performance. The backbone of the model is from BRDNet Tian, C. et al. [28], in which they applied skip connections to add extra information to the high-level features. They used Batch Renormalization whereas this research has used Batch Normalization and added an extra portion of the attention model at the output of the upper and lower CNN.

The model architecture in which an attention module is added at the end of both convolution neural network stacks. The attention module generates a map by utilizing inter-spatial features and extracts essential information to focus on and fine-tune the feature parameters to enhance the results. Table 1 lists the main notations used in this paper.

3.1 *Network Architecture*

The convolutional layer is used to extract features from input images to produce a feature map as shown in Eq. (2). The output of the convolutional layer is obtained by stacking the activation maps of all filters along the depth dimension. Batch Normalization (BN) is used for accelerating the convergence of the model training,

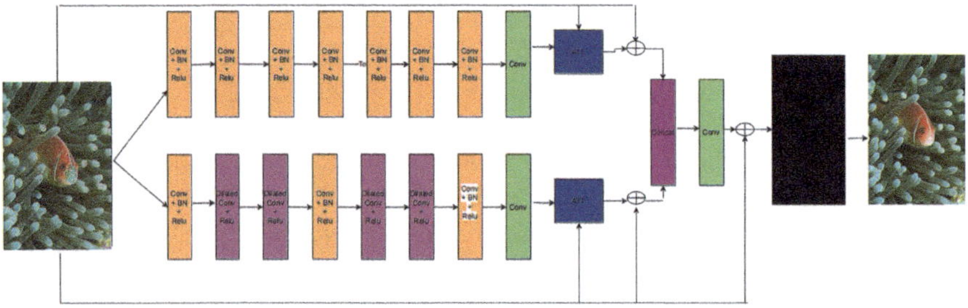

Fig. 1 Proposed model

Table 1 Symbols and notations used

Symbol	Description
x	Original image
y	Noisy image
H	Degradation matrix
e	Additive white Gaussian noise
Fc	Output features of the last convolutional layer
h	Height of CNN
w	Width of CNN
Gp	Global average pooling
d	Dilation rate
Q	Query
K	Key
V	Value
f^{concat}	Feature concatenation
Sa	Self attention
Xn . . . . x_{n-1}	Set of finite features from the attention model

therefore, making the model easy to train. BN is set before the activation function of each layer. Rectified Linear Unit (Relu)—Activation function is used to solve the problem of gradient descent. It is calculated as: f(x) = max (0, x). The dilated convolution is used on the lower network as shown in Eq. (3), as a parameter that can control the dilation rate (d), the convolution kernel is filled with a d-1 number of 0. Lv, Z. et al. [29] expressed dilated convolution as:

$$y = \sum_{i=0}^{k-1} f_i . x - i.d \tag{3}$$

where y and x are the output and input, respectively, d represents the dilation rate. Increasing d or k can increase the range of the receptive field; however, the receptive fields make information obtained by long-distance correlation with the deepening of the network layer leading to loss of the local information. The attention mechanism is the solution to this problem due to its ability to compute correlation. Self Attention (sa) is used to enable the model to focus on important and relevant features more than the insignificant features from the feature maps as shown in Eq. (4). It has three learned linear transformations: query Q, Key K, and values V where Q and K have the same vector dimension dk and V and outputs the same dimension dv. Borrowing from Hamad R. A. et al. [30], Mathematically:

$$f_{sa}^{(hj)}\,(Q, K, V) = softmax\left(\frac{Q.K^T}{\sqrt{d_k}}\right)V \tag{4}$$

where hj is the output from the attention head j, and sa refers to the attention mechanism. Distinct parameters are used in the above Query and value of the attention heads. The outputs from the attention are concatenated and transformed to the dimension of the feature maps. If f_{n} is x_{n} for the attention model n then the output of the attention model n + 1 is x_{n+1}, therefore the concatenated output for the attention models is given by:

$$f^{concat} = \text{concat}\left(x_{n,\ldots\ldots\ldots\ldots,}x_{n-1}\right) \quad (5)$$

4 Experiments

4.1 Datasets

The training dataset is the Waterloo Exploration Database which consists of 3821 images. The models were tested on CBSD68 and Kodak24 datasets. The CBSD68 contained 68 natural images of size 481 × 321. The Kodak24 is composed of 24 natural images of size 500 × 500. For generation of noise images Gaussian noise of sigma = 15, 25, 35, 50, and 75 were added.

4.2 Experimental Setting

The models were trained with sigma values of 15, 25, 35, 50, and 75 being the value of Gaussian noise that was added in the input. The noisy images are the input and the corresponding output are the denoised images. The models were trained on paper space gradient virtual machines with Quadro P4000, 8 CPUs, and 30 GB RAM.

5 Results and Discussion

Peak signal to Noise Ratio (PSNR) has been used as a metric measure for image evaluation. PSNR is referred to as a measurement of the ratio between the maximum power of the signal and the difference between the ground truth and the denoised image, where a higher PSNR indicates closer similarity. It is used to verify the quality resulting from restoration methods Ren, W. et al. [31]. Since its inception about two decades ago, the Structural Similarity (SSIM) measure has established itself as one of the most popular and widely used metrics for evaluating the quality

Table 2 Average PSNR for denoising with sigmas 15, 25, 35, 50, and 75 on the CBSD68 and kodak24 datasets

Datasets	Methods	$\sigma = 15$	$\sigma = 25$	$\sigma = 35$	$\sigma = 50$	$\sigma = 75$
CBSD	CNOATT	31.79	29.29	**	26.36	23.5
CBSD	CWATT	32.7	29.7	27.9	26.8	25.3
Kodak	CNOATT	**	**	**	**	23.9
Kodak	CWATT	33.31	30.6	28.67	27.6	26.1

Table 3 Best PSNR and SSIM for sigma 75

Datasets	Methods	PSNR	SSIM
CBSD	CNOATT	26.53	0.78
CBSD	CWATT	31.27	0.92
Kodak	CNOATT	25.43	0.77
Kodak	CWATT	28.46	0.88

Table 4 Average PSNR and SSIM for sigma 50 for CNOATT, CWATT, CBM3D, DnCNN, and FFDNet

Datasets	Methods	PSNR	SSIM
CBSD	CNOATT	26.36	**
CBSD	CWATT	26.87	0.83
CBSD	CBM3D [33]	27.38	0.76
CBSD	DnCNN [34]	28.01	0.711
CBSD	FFDNet [16]	27.96	0.78

of images and videos, Reznik, Y [32]. This motivated its use in this study to evaluate the models.

To compare the effectiveness of the two models (CNOATT) and the CWATT), the two models were tested on CBSD68 and Kodak24 datasets on sigmas 15, 25, 35, 50, and 75 and testing results are shown in Table 2.

To further demonstrate the effect of introducing an attention model to the output of the individual CNNs (upper and lower) before concatenating, the best PSNR and SSIM were calculated for both models for noise level 75 and the CWATT model performed better than the CNOATT with a difference of 4.74/0.14 for the CBSD68 dataset and 3.03/0.11 for kodak24 dataset for PSNR/SSIM, respectively, as shown in Table 3. Further results for PSNR and SSIM based on sigma 50 are presented in Table 4.

CWATT performed well compared to CNOATT with a difference of 0.51 dB but was outperformed by 0.51 dB,1.14 dB, and 1.09 dB, by CBM3D, DnCNN, and FFDNet, respectively. It outperformed all the state-of-the-art models on SSIM by 0.06, 0.12, and 0.05 for CBM3D, DnCNN, and FFDNet models. Sample noisy and denoised images are presented in Figs. 2 and 3.

Noisy image σ =15 PSNR=24.62 Denoised image PSNR = 30.96

Fig. 2 PSNR for one image from the CBSD68 dataset [35]

Noisy image σ = 15 PSNR= 24.61 Denoised image PSNR = 35.07

Fig. 3 PSNR for one image from Kodak24 dataset [36]

6 Conclusion

In this paper, two CNN models were trained from scratch CNOATT and CWATT and the model structure of the CWATT was changed by adding attention modules at the output of the CNNs and concatenating the features from the attention modules as opposed to concatenating the features without the attention module. The Model with attention performed better than the model without attention and comparably well with state-of-the-art models. CBM3D, DnCNN, and FFDNet models have better PSNR (values got from original papers) than CWATT but CWATT outperformed all three in SSIM. The good performance can be attributed to the fact that all three methods adopt feature extraction without guided selection and therefore error accumulation due to redundant features. SSIM is a better error metric for comparing image quality, Setiadi, D. R. I. M [37]. and Bhatt, R [38]. The proposed model performs better compared to Bhatt, R [38]. Method (A) model by 7.44% and 3.32% for PSNR values for Kodak 24 dataset for sigma values 50 and 75, respectively, and 19.38% and 19.46% for SSIM values for sigmas 50 and 75, respectively. However, it fails to improve the PSNR in comparison with state-of-the-art models, this needs further investigation.

References

1. Zha, Z., Yuan, X., Zhou, J., Zhu, C., & Wen, B. (2020). Image restoration via simultaneous nonlocal self-similarity priors. *IEEE Transactions on Image Processing, 29*, 8561–8576.
2. Sun, S. (2020). Application of fuzzy image restoration in criminal investigation. *Journal of Visual Communication and Image Representation, 71*, 102704.
3. Maru, M., & Parikh, M. C. (2017). Image restoration techniques: A survey. *International Journal of Computer Applications, 160*(6), 15–19.
4. Farhan, R. (2020). Review on some methods used in image restoration. *International Multidisciplinary Research Journal, 10*, 13–16. https://doi.org/10.25081/imrj.2020.v10.6113
5. Khare, C., & Nagwanshi, K. K. (2011). Implementation and analysis of image restoration techniques. *International Journal of Computer Trends and Technology-May to June Issue, 2011*(54), 1–6.
6. Mohammed, O. H., & Mahmood, B. S. (2021). Advance in image and audio restoration and their assessments: A Review.
7. Tian, C., Xu, Y., Li, Z., Zuo, W., Fei, L., & Liu, H. (2020). Attention-guided CNN for image denoising. *Neural Networks, 124*, 117–129.
8. Chen, C., Xiong, Z., Tian, X., & Wu, F. (2018). Deep boosting for image denoising. In *Proceedings of the European conference on computer vision (ECCV)* (pp. 3–18).
9. Tripathi, S., Lipton, Z. C., & Nguyen, T. Q., 2018. Correction by projection: Denoising images with generative adversarial networks. arXiv preprint arXiv:1803.04477.
10. Yue, Z., Zhao, Q., Zhang, L., & Meng, D. (2020). Dual adversarial network: Toward real-world noise removal and noise generation. In *Proceedings of the European Conference on Computer Vision (ECCV)* (pp. 41–58). Springer.
11. Tai, Y., Yang, J., Liu, X., & Xu, C. (2017). Memnet: A persistent memory network for image restoration. In *Proceedings of the IEEE international conference on computer vision* (pp. 4539–4547).
12. Zhang, Y., Tian, Y., Kong, Y., Zhong, B., & Fu, Y. (2021, July 1). Residual dense network for image restoration. *IEEE Transactions on Pattern Analysis and Machine Intelligence, 43*(7), 2480–2495. https://doi.org/10.1109/TPAMI.2020.2968521
13. Mao, X., Shen, C., & Yang, Y. B. (2016). Image restoration using very deep convolutional encoder-decoder networks with symmetric skip connections. *Advances in Neural Information Processing Systems, 29.*

14. Liu, P., Zhang, H., Zhang, K., Lin, L., & Zuo, W. (2018). Multi-level wavelet-CNN for image restoration. In *Proceedings of the IEEE conference on computer vision and pattern recognition workshops* (pp. 773–782).
15. Bhavani, M. D., Murugan, R., & Goel, T. (2022). Robust U-net: Development of robust image enhancement model using modified U-net architecture. *Concurrency and Computation: Practice and Experience, 34*(28), e7347.
16. Zhang, K., Zuo, W., & Zhang, L. (2018). FFDNet: Toward a fast and flexible solution for CNN-based image denoising. *IEEE Transactions on Image Processing, 27*(9), 4608–4622.
17. Yadav, M., & Tiwari, N. (2021). Performance Comparison of Image Restoration techniques using CNN and their applications. In *2021 5th International Conference on Computing Methodologies and Communication (ICCMC)* (pp. 1146–1151). https://doi.org/10.1109/ICCMC51019.2021.9418340
18. Szegedy, C., Liu, W., Jia, Y., Sermanet, P., Reed, S., Anguelov, D., et al. (2015). Going deeper with convolutions. In *Proceedings of the IEEE conference on computer vision and pattern recognition* (pp. 1–9).
19. Szegedy, C., Vanhoucke, V., Ioffe, S., Shlens, J., & Wojna, Z. (2016). Rethinking the inception architecture for computer vision. In *Proceedings of the IEEE conference on computer vision and pattern recognition* (pp. 2818–2826).
20. Srivastava, R. K., Greff, K., & Schmidhuber, J. (2015). Training very deep networks. *Advances in Neural Information Processing Systems, 28.*
21. He, K., Zhang, X., Ren, S., & Sun, J. (2016). Deep residual learning for image recognition. In *Proceedings of the IEEE conference on computer vision and pattern recognition* (pp. 770–778).
22. Targ, S., Almeida, D., & Lyman, K. (2016). Resnet in resnet: Generalizing residual architectures. *arXiv*. https://doi.org/10.48550/arXiv.1603.08029.
23. Lin, T. Y., Dollár, P., Girshick, R., He, K., Hariharan, B., & Belongie, S. (2017). Feature pyramid networks for object detection. In *Proceedings of the IEEE conference on computer vision and pattern recognition* (pp. 2117–2125).
24. Ronneberger, O., Fischer, P., & Brox, T. (2015, October). U-net: Convolutional networks for biomedical image segmentation. In *International conference on medical image computing and computer-assisted intervention* (pp. 234–241). Springer.
25. Vo, X. T., Tran, T. D., Nguyen, D. L., & Jo, K. H. (2021, February). Stair-step feature pyramid networks for object detection. In *International workshop on Frontiers of computer vision* (pp. 168–175). Springer.
26. Li, X., Wang, W., Hu, X., & Yang, J. (2019). Selective kernel networks. In *Proceedings of the IEEE/CVF conference on computer vision and pattern recognition* (pp. 510–519).
27. Anwar, S., & Barnes, N. (2019). Real image denoising with feature attention. In *Proceedings of the IEEE/CVF international conference on computer vision* (pp. 3155–3164).
28. Tian, C., Xu, Y., & Zuo, W. (2020). Image denoising using deep CNN with batch renormalization. *Neural Networks, 121*, 461–473.
29. Lv, Z., Li, J., Dong, C., Li, H., & Xu, Z. (2021). Deep learning in the COVID-19 epidemic: A deep model for urban traffic revitalization index. *Data & Knowledge Engineering, 135*, 101912. https://doi.org/10.1016/j.datak.2021.101912
30. Hamad, R. A., Kimura, M., Yang, L., Woo, W. L., & Wei, B. (2021). Dilated causal convolution with multi-head self attention for sensor human activity recognition. *Neural Computing and Applications, 33*(20), 13705–13722.
31. Ren, W., Liu, S., Zhang, H., Pan, J., Cao, X., & Yang, M. H. (2016, October). Single image dehazing via multi-scale convolutional neural networks. In *European conference on computer vision* (pp. 154–169). Springer.
32. Reznik, Y. (2023). Another look at SSIM image quality metric. *Electronic Imaging, 35*(8), 305–311.
33. Dabov, K., Foi, A., Katkovnik, V., & Egiazarian, K. (2007). Image denoising by sparse 3-D transform-domain collaborative filtering. *IEEE Transactions on Image Processing, 16*(8), 2080–2095.

34. Zhang, K., Zuo, W., Chen, Y., Meng, D., & Zhang, L. (2017). Beyond a gaussian denoiser: Residual learning of deep cnn for image denoising. *IEEE Transactions on Image Processing, 26*(7), 3142–3155.
35. Michele, C. (2018). CBSD68 dataset. https://github.com/clausmichele/CBSD68-dataset/tree/master/CBSD68
36. Franzen, R. (1999). Kodak lossless true color image suite. 4(2), 9. http://r0k.us/graphics/kodak
37. Setiadi, D. R. I. M. (2021). PSNR vs SSIM: Imperceptibility quality assessment for image steganography. *Multimedia Tools and Applications, 80*(6), 8423–8444.
38. Bhatt, R., Naik, N., & Subramanian, V. K. (2021). SSIM compliant modeling framework with denoising and deblurring applications. *IEEE Transactions on Image Processing, 30*, 2611–2626.

A Contribution to the Development of Sustainable Target Value Streams with Machine Learning Considering Material Flow Cost

Mick Geisthardt and Lutz Engel

1 Introduction

Value stream mapping is considered as a central element of lean management, which is employed globally for process improvement within industrial production. The objective is to maximize value creation and eliminate waste, which consequently requires consideration of all types of waste. Due to existing limitations to waste types assessable via lead time, material and energy-based wastes remain substantially unconsidered, resulting in incomplete process improvements solely favoring the lead time.

Especially resource efficiency gains increasingly important for companies involved in industrial production. Recent price trends as well as their forecasted progression implicate that material and energy costs will continue to rise significantly in the future. As for example, according to the German Federal Association of Energy and Water Industries, the average electricity price for new contracts in German industry including electricity tax has almost doubled over the last 10 years (14.33 ct/kWh in 2012, 26.64 ct/kWh in January 2022) [1].

To integrate the aspect of resource efficiency and the associated cost factor within lean management, existing research has extended value stream mapping to include the concept of material flow cost accounting in accordance with DIN EN ISO 14051:2011 [2]. This utility-oriented extension resolves the lead time factors exclusivity and enables the initially neglected consideration of material and energy-based wastes. Concerning applicability and functionality, this extended value stream mapping entails significant expenses in terms of data acquisition and processing, as well as calculation complexity and time-cost balance.

M. Geisthardt (✉) · L. Engel
Jade University of Applied Sciences, Wilhelmshaven, Germany
e-mail: mick.geisthardt@jade-hs.de; lutz.engel@jade-hs.de

J. Marx Gómez et al. (eds.), *Artificial Intelligence Tools and Applications in Embedded and Mobile Systems*, Progress in IS,
https://doi.org/10.1007/978-3-031-56576-2_19

To enable value-adding utilization of rising volume and complexity of data for generation of new target value streams in the context of industrial progress, the necessity to develop a model concept for a digital application arises. Furthermore, it must be addressed, which influence the new insights from value stream mapping extended by material flow cost accounting have on the overall improvement. This is part of this research paper, which is described here.

Initially, the background of value stream mapping and its advancement in context of industrial production is examined. Subsequently, a machine learning-based model concept for discovery of sustainable target value streams is developed. Focusing on possible components of systematic development, potential of material flow costs utilization, and prospective application. The digital application based on the presented model concept is expected to assist improvement teams and contribute to the discovery of waste-optimized and more sustainable target value streams in an industrial environment.

2 Related Works

For more than 20 years, value stream mapping has been used within industrial production to identify improvement potentials and address them by designing target value streams. The purpose is the realization of value creation without waste. Initially, the current value stream is captured. Subsequently, the three Ms., "Muda," "Mura," and "Muri" are applied to identify possible wastes occurring within the value stream [3]. "Muda" refers to the eight wastes, such as transport, inventory, unnecessary movement, or overproduction. "Mura" indicates wastes due to uneven utilization of capacities and "Muri" detects overloads of capacities. All these wastes are wastes that are assessable via lead time. These wastes are visually highlighted in the current value stream by an improvement team and evaluated in terms of time. No monetary evaluation of these wastes is conducted.

Afterward, a target value stream is designed by the same improvement team. It is derived from the current value stream but optimized in terms of lead time by eliminating identified waste and applying guidelines. Guidelines according to Rother [4] are, e.g., production according to customer takt time, continuous flow production or supermarket pull principle. Guidelines according to Erlach [5] are, e.g., bottleneck control or separation of production process and material flow. Aimed at and guiding is the ideal state of the Toyota Production System consisting of 0 defects, 100% value creation, one piece flow (in sequence on demand), and safety for people [6]. Thus, when designing the target value stream, mainly concerning the flow of one product family through the production processes, the value stream and thus the material and information flows are improved. The result is a continuous flow with low inventories and short lead times, which naturally also reduces costs.

Due to the limitation to waste types that are assessable via lead time, material, and energy-based wastes, such as excessive energy consumption or wastage of material, remain substantially unconsidered throughout the value stream. To resolve

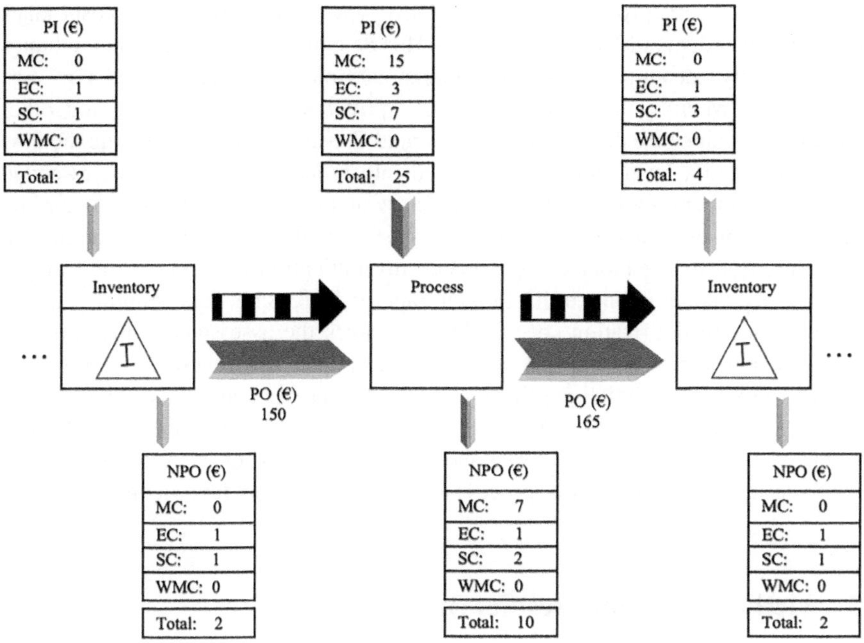

Fig. 1 Representation of material flow costs within extended value stream mapping

the resulting aspect of incomplete process improvement solely favoring the lead time, value stream mapping has been extended by the concept of material flow cost accounting [2].

This extended form of value stream mapping incorporates terminology from material flow cost accounting in accordance with DIN EN ISO 14051:2011 [7], whereby the defined concept of a quantity center is equated with a process or inventory within the value stream. As illustrated in Fig. 1, the input of a process is differentiated as product input (PI) and the output of a process as product output (PO) and non-product output (NPO). While the product output represents those costs that increase the product value from the customer's perspective, the non-product output includes those costs that do not increase the value of the product. Accordingly, product output is associated with value creation and non-product output with waste.

As within material flow cost accounting, the costs for product input, product output, and non-product output are sub-divided into material costs (MC), energy costs (EC), system costs (SC), and waste management costs (WMC). The representation of these cost components within the fundamentally flow-oriented value stream mapping is conducted in accordance with DIN EN ISO 14051:2011 as a sankey diagram, whereby the width of a sankey arrow represents the cost magnitude and the color represents the respective cost type.

Since not every process contains all cost types, the allocation of costs takes place within a demand-based calculation. This demand-based calculation facilitates comparability between values from value stream mapping and material flow cost accounting, whereby the costs of services provided, material flows, and other quantitative values are determined on a demand-dependent basis (following the flow principle) at unit level and then converted into monetary values. Thus, extended value stream mapping provides both monetary and temporal evaluations regarding value creation and waste at the level of a process as well as of the entire value stream.

In addition to the methodology-related diversification of the input and output metrics, the value-adding utilization of rapidly increasing data volumes due to automation and digitization specifically challenges the assessment capabilities of improvement teams. Within this context, for example, reinforcement learning can be utilized within a digital application to assist improvement teams in identifying target value streams [8]. Thereby existing research continues to focus on traditional forms of waste, neglecting material and energy-based wastes even in 2022, which continues to lead to incomplete process improvements solely favoring the lead time.

3 Methodology and Results Discussions

To digitally enable the availability of both traditional product and process data and, prospectively, resource data from material flow cost accounting, further research on value stream mapping has resulted in the ability to analyze the information logistics in the direct and indirect areas [9]. Thereby, the process notation allows to capture and evaluate the information flows of activities and key performance indicators (KPIs) via storage media and its usage. Media breaks between storage media become visible, allowing targeted optimization projects to be initiated as part of the holistic digitization process. As a result of the extensive implementation of digital technologies within the production system, the availability of data in the form of KPIs and information is maximized. The possibility of instrumental or conceptual utilization of these KPIs and information via a digital application emerges.

In the context of this research paper, there is the possibility to utilize the KPIs specified in Table 1 as a basis for the featured model concept.Established traditional

Table 1 Selected traditional and extended value stream mapping KPIs

Traditional value stream mapping KPIs	Extended value stream mapping KPIs
Lead time	Product input (PI)
Cycle time (C/T)	Product output (PO)
Changeover time (C/O)	Non-product output (NPO)
Uptime	Absolute monetary value creation
Overall equipment effectiveness (OEE)	Relative monetary value creation
Scrap rate	Degree of value creation

KPIs are known from existing research and industrial practice. Extended KPIs are derived from value stream mapping extended by material flow cost accounting and allow the consideration of the resource aspect on process and value stream level through product input, product output, and non-product output.

To measure the monetary value creation within a process, the absolute monetary value creation indicates the value created in a process by the difference between the product output of a process and the product output of the corresponding upstream process (see Eq. 1).

$$Absolute\ Monetary\ Value\ Creation = Costs_{PO} - Costs_{PO\ (Upstream)} \tag{1}$$

The relative monetary value creation indicates the ratio of the absolute monetary value creation of a process to the product input. It is, therefore, a monetary assessment of the efficiency of a process (see Eq. 2).

$$Relative\ Monetary\ Value\ Creation = Costs_{PO} - Costs_{PO\ (Upstream)}/Costs_{PI} \tag{2}$$

The degree of value creation can be applied as a comparative monetary indicator for the entire process chain of a value stream. Thereby, the total costs of the product output are set in relation to the total costs of the product input (see Eq. 3).

$$Degree\ of\ Value\ Creation = Costs_{PO\ (Total)}/Costs_{PI\ (Total)} \tag{3}$$

For the value-adding utilization of the described KPIs via a digital application, three basic components are necessary for the systematic development of the model concept (see Fig. 2).

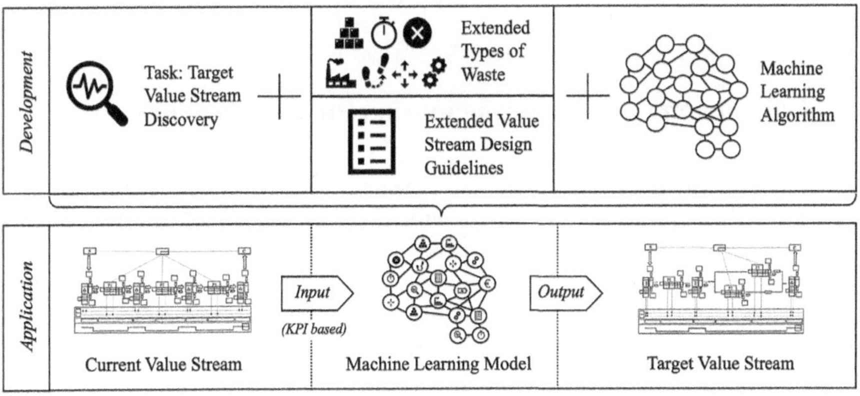

Fig. 2 Concept for sustainable target value stream discovery via machine learning

The first component of the model concept is the concrete task of the digital application. Within this concept, the task is to assist improvement teams in discovering waste-optimized and more sustainable target value streams toward the ideal state within an industrial environment. Focus is thereby on the methodical transformation process from the KPI-based current value stream to the target value stream.

The second component is the rules according to which the digital application is supposed to solve the specified task, while at the same time being able to learn. These rules are the traditional [3] and extended types of waste for identifying potentials within the KPI-based current value stream. Additionally, the traditional [4, 5] and extended guidelines serve as rules for the transformation process from the current value stream to the target value stream.

The last component of the model concept is the machine learning algorithm required for analysis of KPIs-based current value stream, application of rules, and the solution of the task. Within the development of digital applications, algorithms of the three main forms of machine learning can be fundamentally utilized. These include supervised learning in which learning is based on input-output pairs, unsupervised learning in which learning is based on patterns in the input without explicit feedback, and reinforcement learning in which learning is based on rewards and punishments [10].

Successful synthesis of these three components and the associated realization of the outlined model concept results in a digital application that leverages machine learning. Within the scope of its prospective application, this application can take a KPI-based current value stream as input, perform value-adding analysis, and assist in the discovery of target value streams through the resulting output. Due to the characteristics of the machine learning algorithm, the digital application learns continuously, which in turn continuously improves the output quality.

Overall, this digital application to be developed can thus assist improvement teams and can contribute to discovering waste-optimized and more sustainable target value streams in an industrial environment.

4 Conclusion and Recommendation for Future Work

As industrial progress continues, the concept of value stream mapping, which originates from lean management, continues to exhibit high potential when it comes to maximizing value creation and eliminating waste. Extending value stream mapping by material flow cost accounting can thereby enable to factor of material and energy-based wastes and thus relativize the existing lead time exclusivity. Within the value-adding utilization of increasing volumes of data, it seems distinct that a digital application will be necessary in the future.

To address such a digital application, the hypothesis of this research paper is a systematic development of a model concept to discover waste optimized and more sustainable target value streams toward the ideal state. In doing so, the complementary potentials resulting from material flow cost accounting and machine learning are highlighted. Through their prospective application, a contribution can be made to assist improvement teams in their work within an industrial environment.

In continuation of this research paper, the influence of insights gained from value stream mapping extended by material flow cost accounting on the improvement is to be investigated in greater detail. The results of this investigation are to be utilized to expand the existing types of waste and guidelines and to re-evaluate their weighting. At the same time, the traditional ideal state must be questioned and, if necessary, updated, e.g., to include 0 emissions. The corresponding model concept for a digital application should then be entirely developed and evaluated. Thus, this research project provides a contribution to the development of sustainable target value streams with machine learning considering material flow costs.

This research project will be funded by the Jade University of Applied Sciences Doctoral Scholarship Program in cooperation with the Carl von Ossietzky University of Oldenburg and Prof. Dr.-Ing. habil. Jorge Marx Gómez.

References

1. Bundesverband der Energie- und Wasserwirtschaft e.V. Strompreisanalyse Juli (2022). Retrieved August 24, 2022, from https://www.bdew.de/service/daten-und-grafiken/bdew-strompreisanalyse/
2. Engel, L., & Kranhold, M. (2021). Lean Management wird zum Green Management. *Um Materialflusskosten erweiterte Wertstromanalyse. Deutsches Ingenieurblatt, 2021*(6), 1–6.
3. Ohno, T. (2013). *Das Toyota-Produktionssystem* (3rd ed.). Campus Verlag GmbH.
4. Rother, M., & Shook, J. (2011). *Sehen lernen. Mit Wertstromdesign die Wertschöpfung erhöhen und Verschwendung beseitigen. Version 1.4*. Lean Management Institut GmbH (Workbooks für Lean Management).
5. Erlach, K. (2010). *Wertstromdesign. Der Weg zur schlanken Fabrik* (2nd ed.). Springer-Verlag GmbH.
6. Rother, M. (2009). *Toyota Kata: Managing people for improvement, adaptiveness and superior results*. McGraw-Hill Global Education Holdings LLC.
7. Deutsches Institut für Normung e.V (Ed.). (2011). *DIN EN ISO 14051:2011-12. Umweltmanagement—Materialflusskostenrechnung—Allgemeine Rahmenbedingungen (ISO 14051:2011)*. Beuth Verlag GmbH.
8. Langlotz, P., Klar, M., Glatt, M., & Aurich, J. C. (2022). Automatisierte Wertstrommethode unter Nutzung von Reinforcement Learning. Ein Konzept zur Identifikation und Nutzung von Verbesserungspotentialen. *Zeitschrift für wirtschaftlichen Fabrikbetrieb, 117*(6), 395–399.
9. Meudt, T., Metternich, J., & Abele, E. (2017). Value stream mapping 4.0: Holistic examination of value stream and information logistics in production. *CIRP Annals, 66*(1), 413–416.
10. Russell, S., & Norvig, P. (2021). *Artificial intelligence. A modern approach* (4th ed.). Pearson Education Inc..

Mask R-CNN Model for Banana Diseases Segmentation

Christian A. Elinisa and Neema Mduma

1 Introduction

Over 70 million Africans depend on bananas for their food security and for generating revenue. Most of the bananas produced in Africa are consumed locally. Uganda, Rwanda, and Burundi in East Africa have the highest per capita consumption rate in the world [1]. Green bananas are the primary source of carbohydrates for around 30% of Tanzanians. Tanzania's principal banana-producing regions include Kagera, Kilimanjaro, Mbeya, Kigoma, Arusha, Tanga, and Morogoro [2]. Bananas provide a sustainable source of food supply because they produce fruits around the year. Bananas can be consumed fresh as fruits, cooked, fried, and processed to make juice, beer, and baby food [1]. Bananas are cheap compared to citrus fruits and apples, and they provide energy, vitamins, and potassium, making them the most preferred fruit [3]. However, banana production is highly affected by fungal diseases like Fusarium Wilt Race 1 and Black Sigatoka which causes high losses in yields.

Fusarium Wilt Race 1 is a harmful banana disease globally caused by a soil-borne fungus called *Fusarium oxysporum f.* sp. cubense (Foc). The fungus starts harming the plant through the roots by blocking the passage of water and nutrients. This causes leaf discoloration to yellow and wilting. The leaves eventually collapse around the pseudostem and fall dry. The fungus also causes the splitting of the pseudostem base. The disease cannot be eradicated using chemical pesticides, but farmers can overcome it by replacing susceptible cultivars with resistant ones in affected soils or shifting their production to unaffected land [4].

C. A. Elinisa (✉) · N. Mduma
School of Computational and Communication Science and Engineering, The Nelson Mandela African Institution of Science and Technology, Arusha, Tanzania
e-mail: elinisac@nm-aist.ac.tz

J. Marx Gómez et al. (eds.), *Artificial Intelligence Tools and Applications in Embedded and Mobile Systems*, Progress in IS,
https://doi.org/10.1007/978-3-031-56576-2_20

Black Leaf Steak Disease (BLSD), also called Black Sigatoka, is a disease that causes leaf spots on banana plants. *Pseudocercospora fijiensis*, formerly known as *Mycosphaerella fijiensis*, is the air-borne fungus that causes this leaf spot disease. The disease interferes with photosynthesis by consuming the banana leaves. Small specks that later develop into streaks parallel to leaf veins are the first signs of the disease for 14–20 days after infection. The streaks combine to produce spots that eventually merge into a chlorotic halo and result in broad necrosis. Farmers can apply fungicides weekly to control the disease [5].

Farmers face the challenge of detecting plant diseases early because they use the traditional simple naked-eye observation method. This method is cumbersome, time consuming, and also prone to errors, especially on large farms. However, this method can be automated through the use of computer vision and deep learning to minimize the mentioned shortcomings. Deep learning has advanced and yielded outstanding results in plant disease detection. It can be used for identifying, classifying, segmenting, detecting, and quantifying plants and their diseases. Inspired by this growth, this research proposes a Mask R-CNN model for banana disease segmentation. The model localized the affected area on the banana plant by the disease in the image. Farmers will be able to identify banana diseases earlier and take appropriate measures to reduce losses and increase production.

2 Related Works

Several studies have applied deep learning and computer vision to segment and detect plants and their diseases. Many studies have used Convolution Neural Networks for crop disease classification while others have used Mask R-CNN, for instance, segmentation.

Researchers in [6] proposed a plant leaf disease detection method that combined four CNN models. The CNN models used were Inception, Densenet, Inception Resnet, and Resnet. The study used an open-source dataset of 36,258 images for 10 plant species with 61 classes, including healthy and diseased plant leaves. In this study, the dataset was divided into 31,718 training images and 4540 validation images. Compared to the outcomes of using a single CNN model, a stacking method was used to process the results of the four CNN models and achieved an accuracy of 87%, which was significantly higher.

Furthermore, researchers in [7] suggested a mobile application that uses deep learning for the early detection of Fusarium Wilt Race 1 and Black Sigatoka banana diseases. Their experiment used a dataset of 3000 banana leaves. The study pre-trained Inceptionv3 and Resnet152 CNN architectures. Inceptionv3 achieved an accuracy of 95.41% and Resnet152 achieved an accuracy of 99.2%. Due to lower memory requirements, the Inceptionv3 model was deployed in an Android mobile application. The mobile application was reported to detect the two diseases with a high confidence level.

Moreover, researchers in [8] proposed a poly-house solution for plant monitoring and early disease detection using CNN. Their dataset included images of corn, strawberry, grape, tomato, and potato leaves containing healthy and diseased leaves. The Convolutional Neural Network model was trained to detect plant diseases with a 0.25 negligible loss during training. The model achieved a prediction accuracy of 85%.

Also, researchers in [9] suggested a CNN model with an optimized activation function for plant disease detection and classification. The researchers developed a new mathematical activation function and compared it to existing activation functions like ReLu. In an experiment, their activation function improved the CNN model's accuracy and performance by achieving an accuracy of 95% and improving training speed by 83%. The study also calculated the affected area by the disease using K-means clustering.

Even though the discussed studies have addressed the plant disease detection problem using Convolutional Neural Networks deep learning models, to the best of our knowledge we have noticed that only a small number of studies have focused on developing models that can segment the infected area of the plant leaf or stalk image.

Researchers in [10] proposed an early solution for detecting the effects of Tuta Absoluta on tomato plants. To segment the effects of Tuta absoluta on tomato leaf images, the study proposed using semantic and instance segmentation models based on U-Net and Mask R-CNN deep convolutional neural networks. A total of 5235 tomato leaf images were used, categorized into healthy and infested by Tuta absoluta. The Mask R-CNN model had a mean average precision of 85.67%, while the U-Net model had an intersection over union of 78.60% and a dice coefficient of 82.86%.

Furthermore, researchers in [11] suggested a model that successfully conducts instance segmentation for seven strawberry diseases using the Mask R-CNN architecture. The study used a dataset of 2500 images for the seven diseases. ResNet backbone and data augmentation were used with the Mask R-CNN model to segment the diseases. The model achieved a mAP of 82.43%.

Moreover, researchers in [12] proposed a tomato disease identification and detection technique based on deep convolutional neural networks and object detection algorithms. The study used the Faster R-CNN model for the identification of tomato diseases and the Mask R-CNN model for the detection and segmentation of the affected areas, including location and shapes. This study used a dataset of 286 images from the internet (that was expanded to 1430 images) with 10 classes of tomato diseases and 1 healthy class. A mean Average Precision of 88.53% and 99.64% was achieved for Faster R-CNN and Mask R-CNN, respectively.

In general, most of the conducted studies presented models that showed outstanding performances. However, none of the presented studies addressed the segmentation of banana diseases. Therefore, this research proposes a Mask R-CNN model for banana disease segmentation using field-collected images.

3 Materials and Methods

3.1 The Dataset

The dataset used in this study was collected in the Arusha, Kagera, Kilimanjaro, Mbeya, and Dar es Salaam regions in Tanzania. The selection of the study areas was based on the consideration of banana production. Pictures of healthy and diseased banana leaves and stalks were taken using mobile phones and labeled with the guidance of agricultural experts and plant pathologists. The dataset was divided into three classes, which are images of healthy banana leaves; images of banana leaves affected by Black Sigatoka disease; and images of banana leaves and stalks affected by Fusarium Wilt Race 1. The dataset was divided into a training set and a validation set in a ratio of 80:20. Table 1 shows the distribution of the dataset, and Fig. 1 shows an example of a healthy banana leaf, a banana leaf affected by Black Sigatoka disease, and a banana leaf and stalk affected by Fusarium Wilt Race 1 disease.

3.2 Data Preprocessing

Data preprocessing in this work included data cleaning, renaming, and data annotation.

Table 1 Dataset distribution

Banana images	Training set	Validation set	Total images
Healthy leaves	1600	400	2000
Leaves affected by black Sigatoka	1600	400	2000
Leaves and stalks affected by fusarium wilt race 1	1600	400	2000
Total	4800	1200	6000

Fig. 1 Examples from the dataset

Data Cleaning. The dataset was collected and saved with some duplicates. VisiPics, a free software, was used to remove 3280 duplicates from the images. This software was our default choice because it is open source and easy to use. Only strictly similar (or identical) images were removed. The dataset presented in Table 1 contains clean images recorded after removing duplicates.

Data Renaming. For simplicity, the clean images in each class were renamed to comprise image numbers. For example, for healthy images, the first one was HEALTHY_1.jpg and the numbers kept on increasing to HEALTHY_2000.jpg. We used Bulk Rename Utility free software to rename images for all classes. This tool was also used because it is open source and simple to use.

Data Annotation. The 6000 images for all classes were annotated for the instance segmentation task. Mask R-CNN requires the images to have masks with labels in order to obtain accurate predictions of banana-infected areas and their shapes. LabelMe which is an open-source software was used to annotate the images to be used in instance segmentation by Mask R-CNN. The specific procedure was to manually draw a mask around the regions of interest in the banana plant image using irregular polygons and then label them with the class name. For example, for banana leaf images affected by Black Sigatoka disease, an irregular polygon was drawn around each spot showing the damage of the disease on the leaf, and each polygon was given the label "Black Sigatoka." The irregular polygons drawn around the spots that showed the damage of Fusarium Wilt Race 1 disease on banana leaves and stalks were given the label "fusarium wilt" For the healthy banana leaves, an irregular polygon was drawn around the whole leaf area or around all the healthy leaves in the image, and the polygons were given the label "healthy." There was at least one irregular polygon for each image in the dataset. Each image file had its corresponding annotation file in the same folder with the same name except for the extension. LabelMe saves its annotations in JSON format. Figure 2 shows how the dataset was annotated using LabelMe software.

3.3 Proposed Models

We focused on developing an instance segmentation deep learning model for detecting and segmenting banana diseases using Mask R-CNN. Mask Region-based CNN is an extension of Faster R-CNN, whereby it adds a segmentation mask predicting branches on each region of interest in parallel with the already existing classification and bounding box regression branch [13]. Transfer learning was applied by using the latest Mask R-CNN trained weights from the COCO dataset. Also, a Convolutional Neural Network model was used to classify these diseases. The CNN model performs classification of the banana diseases while the Mask R-CNN model performs classification and adds a bounding box and a segmentation mask in the area of interest (i.e., the area on the banana leaf affected by the diseases).

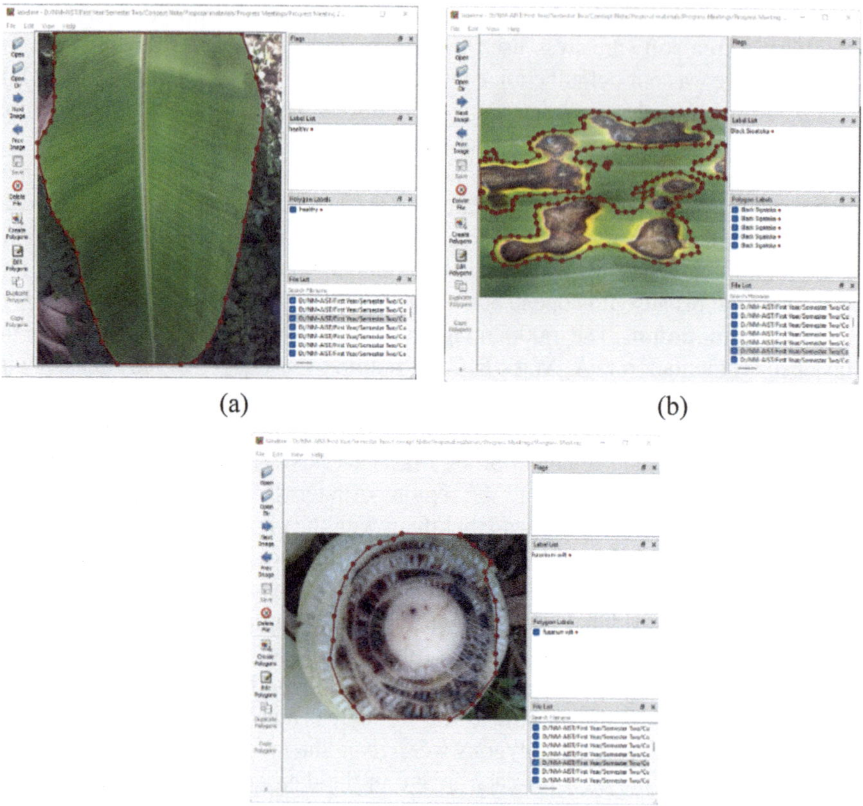

(a) (b)

(c)

Fig. 2 How LabelMe was used to manually annotate images. (**a**) Annotation of a healthy banana leaf image; (**b**) Annotation of a banana leaf image affected by Black Sigatoka disease; and (**c**) Annotation of a banana stalk image affected by Fusarium Wilt Race 1 disease

3.4 Model Training

Mask R-CNN: Hyper-parameter tuning. In our experiment, the ResNet101 backbone architecture was used in the Mask R-CNN model. We used a batch size of 1 since we trained on 1 GPU and 1 image per GPU. We set the detection minimum confidence at 0.9, which keeps the regions of interest with a higher score than this threshold and skips those below. The learning rate was set to 0.001, learning momentum to 0.9, and weight decay to 0.0001. We trained over 41 epochs and evaluated the model on the validation set at the end of each epoch.

CNN: Hyper-parameter tuning. In our implementation, we used a sequential model. We defined four convolutional layers, each followed by a max pooling layer. All the convolutional layers had 64 filters, a kernel size of 3, a stride of 2, padding was set to same, and they used the ReLu activation function. These were followed

Table 2 Hyperparameters used for training the CNN model

Parameters	Value
Batch size	64
Epoch	50
Optimizer	Adam
Loss	Categorical Crossentropy
Metric	Accuracy, precision, recall and F-measure

by a flattened layer, which was followed by a dense layer with 512 neurons with the ReLu activation function. The output dense layer had three neurons, which was the number of our classes, and had a softmax activation function. Our images were rescaled by 1/255 to the range of (0–1) and they were resized to 512 x 512 pixels. To expand our training dataset, data augmentation was used, which included a random image rotation up to 40 degrees left or right; image translation by up to 20% horizontally or vertically; image shearing by up to 20%; image zooming by up to 20%; random image flipping vertically or horizontally; and using the nearest neighbors to fill up any missing pixels following a move or shear. Table 2 summarizes the hyperparameters used to train the CNN model.

3.5 Experiment Setting

Our experiments were done on a computer having Windows 11 Pro with one Intel(R) Core(TM) i5-8250U CPU @ 1.60GHz 1.80 GHz with 8GB RAM. We run our notebook on Google Colab Pro with a Tesla P100-PCIE GPU having 27.3GB of RAM. We used Python 3, Keras library, and TensorFlow backend to implement our network.

4 Results and Discussion

4.1 Training Time

Training time is a crucial measure of a model's performance. The training times of our model's experiments are illustrated in Table 3. Results showed that the CNN model with augmentation took 164.91 min longer than the CNN model without augmentation because of all the image processing. Furthermore, the Mask R-CNN model with ResNet101 takes a longer time to train than the CNN models because it is much more complex in structure compared to the CNN models.

Table 3 Model training time

Model	Training time (Mins)
CNN	120.72
CNN with augmentation	285.63
Mask R-CNN—ResNet101	1166.52

Table 4 CNN model performance for detecting banana diseases

Model	Epoch	Validation loss	F-measure (%)	Precision (%)	Recall	Validation accuracy (%)
CNN	50	0.2396	50.55	33.86	1	96.75
CNN with augmentation	50	0.0898	49.99	33.33	1	96.75

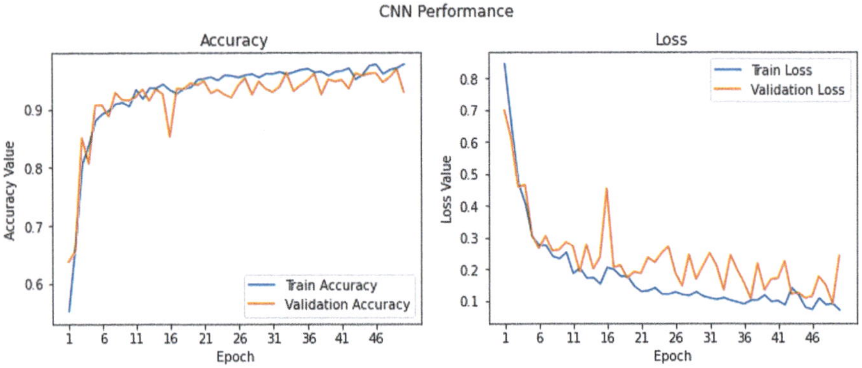

Fig. 3 Graphs showing CNN model performance. A graph of accuracy over Epoch on the left and a graph of loss over Epoch on the right

4.2 *CNN Results and Discussion*

The results of the CNN models are summarized in Table 4. These results were selected after performing several experiments and obtaining the best results. Results showed that the CNN model with and without augmentation achieved a validation accuracy of 96.75% and 96.75% and a validation loss of 0.0898 and 0.2396, respectively. Figure 3 shows the graphs of the CNN model's performance. On the accuracy over epoch graph on the left of Fig. 3, results show that the validation accuracy rose rapidly to the sixth epoch, then increased slowly with small fluctuations, hitting a maximum of 0.9675, while the training accuracy followed the same pattern without fluctuations, hitting a maximum of 0.97. This shows that the model could generalize well without overfitting. On the loss over epoch graph on the right of Fig. 3, results showed that the training loss decreases rapidly in the early epochs, and at epoch 21, it starts to decrease steadily until the end, while the validation loss decreases rapidly from the beginning, and from epoch 6, it started decreasing steadily with

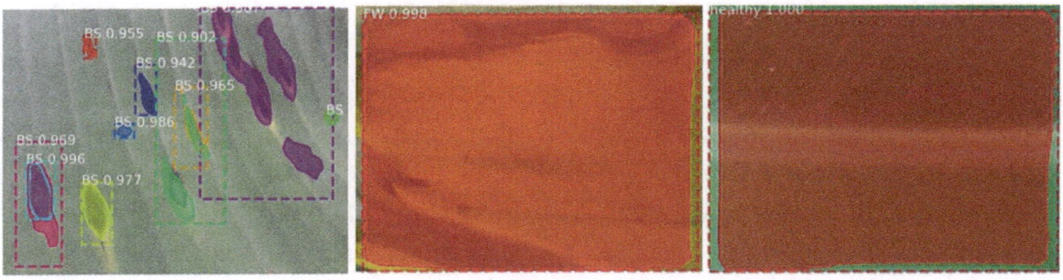

(a) (b) (c)

Fig. 4 Examples of segmentation predictions made by the proposed Mask R-CNN model. (**a**) Image segmentation of a leaf affected by Black Sigatoka disease, (**b**) image segmentation of a leaf affected by Fusarium Wilt Race 1 disease, and (**c**) image segmentation of a healthy leaf

fluctuations. This suggests that both early in the training process and later on, the model fitted well with the features of our dataset.

The achieved CNN model accuracy of 96.75% is considered good since a good CNN model should achieve an accuracy of at least 70% [14]. Similar results were obtained by [7] when deploying an Inceptionv3 model which achieved an accuracy of 95.41%. As a result, the proposed model is accurate and suitable for use in the early detection of banana diseases.

4.3 Mask R-CNN Results and Discussion

The Mask R-CNN model achieved a mean Average Precision (mAP) of 0.04529. The model was able to accurately segment the affected areas of the banana plant by Black Sigatoka and Fusarium Wilt Race 1 diseases and the healthy banana leaves with high confidence scores. Figure 4 illustrates examples of how the proposed Mask R-CNN model predicted the segmentations of the banana leaves affected by Black Sigatoka disease, banana leaves affected by Fusarium Wilt Race 1 disease, and healthy banana leaves, respectively.

5 Conclusion

This study proposes a Mask R-CNN architecture for banana diseases segmentation. The work also includes a CNN model for classifying banana diseases. The banana diseases involved in this work are Black Sigatoka and Fusarium Wilt Race 1. The results of our experiment showed that the Mask R-CNN ResNet101 model achieved a mAP of 0.04529 in segmenting banana diseases, while the CNN model achieved an accuracy of 96.75% in classifying the diseases. Furthermore, the Mask R-CNN model was able to accurately segment areas where the banana leaves and stalk were damaged by Black Sigatoka and Fusarium Wilt Race 1 diseases in the image dataset. This study shows that deep learning techniques have the potential to detect banana diseases and other plant diseases early enough so that farmers can take the required measures to control or minimize the harmful effects of these diseases and rescue their yields. We intend to explore other segmentation models and deploy the best model in a mobile application to enhance usability in the future. This mobile application will help farmers, extension officers, and plant pathologists in the early detection of banana diseases and provide farmers with recommendations on how to control these diseases so as to rescue their plants and increase production.

Disclosure Statement

No conflicts of interest exist, according to the authors, with the publication of this paper.

Funding This study is supported by Data Science Africa (DSA) and the International Development Research Centre (IDRC) and the Swedish International Development Cooperation Agency (SIDA) through Lacuna Fund in Agriculture and the African Centre for Technology Studies (ACTS).

References

1. Daniel. (2016, February 8). Bananas from Africa. [Online]. http://inafrica24.com/modernity/bananas-from-africa/
2. Suleiman, R. (2018). Local and regional variations in conditions for agriculture and food security in Tanzania: A review.
3. Voora, V., Larrea, C., & Bermudez, S. (2020). Global market report: Bananas.
4. Vézina, A. (2022, March 9). Fusarium wilt of banana. [Online]. https://www.promusa.org/Fusarium+wilt
5. Vézina, A., & Van den Bergh, I. (2020, August 5). Black leaf streak. [Online]. https://www.promusa.org/Black+leaf+streak.
6. Guan, X. (2021). A novel method of plant leaf disease detection based on deep learning and convolutional neural network. In *2021 6th International Conference on Intelligent Computing and Signal Processing (ICSP).*
7. Sanga, S., Mero, V., Machuve, D., & Mwanganda, D. (2020). Mobile-based deep learning models for banana diseases detection. In *Computer vision and pattern recognition.*
8. Radha, N., & Swathika, R. (2021). A Polyhouse: Plant monitoring and diseases detection using CNN. In *2021 International Conference on Artificial Intelligence and Smart Systems (ICAIS).*
9. Yadhav, S. Y., Senthilkumar, T., Jayanthy, S., & Kovilpillai, J. J. A. (2020). Plant disease detection and classification using CNN model with optimized activation function. In *2020 International Conference on Electronics and Sustainable Communication Systems (ICESC).*
10. Loyani, L. K., Machuve, D., & Bradshaw, K. (2021). Segmentation of Tuta Absoluta's damage on tomato plants: A computer vision approach. *Applied Artificial Intelligence, 35*(14), 1107–1127.
11. Afzaal, U., Bhattarai, B., Pandeya, Y. R., & Joonwhoan, L. (2021). An instance segmentation model for strawberry diseases based on mask R-CNN. *Sensors, 21*(19), 6565.

12. Wang, Q., Qi, F., Sun, M., Qu, J., & Xue, J. (2019). Identification of tomato disease types and detection of infected areas based on deep convolutional neural networks and object detection techniques. *Computational Intelligence and Neuroscience, 2019*, 9142753.
13. He, K., Gkioxari, G., Doll'ar, P., & Girshick, R. (2020, February 01). Mask R-CNN. *IEEE Transactions on Pattern Analysis and Machine Intelligence, 42*(2), 386–397.
14. Maxwell, A. E., Warner, T. A., & Guillén, L. A. (2021, June 23). Accuracy assessment in convolutional neural network-based deep learning remote sensing studies—Part 1: Literature review. *Artificial Intelligence and Machine Learning with Applications in Remote Sensing, 13*(13), 2450.

The Artificial Neural Network-Based Smart Number Plate for Vehicles with Real-Time Traffic Signs Recognition and Notification

Alexandre Niyomugaba, Neema Mduma, and Kisangiri Michael

1 Introduction

A vehicle's number plate is an essential component that plays a vital role in differentiating the vehicle from others. The vehicle's license plates contain unique numbers and letters that support the dedicated authorities like traffic police and national revenues to easily identify the vehicle's information, such as the vehicle's owners, state, and country of origin. With the use of Artificial Neural Networks (ANN) [1] and the Internet of Things (IoT) [2], technology is fast penetrating every area, including modern transportation such as automated vehicle identification [3], real-time vehicle's speed detection and number plate recognition [4]. However, even though these systems have been implemented, some challenges still exist in traffic monitoring and control, such as the lack of real-time driving assistance that can support drivers to obey traffic signs and the lack of a real-time accident locator once it happens. Some drivers obey traffic signs only when reaching where those smart cameras have been fixed.

Therefore, this paper describes the development of a Smart Number Plate (SNP) for vehicles to assist drivers in recognizing earlier and identifying the traffic signs, knowing the vehicle's location and speed while driving, and notifying the police station once any accident happens or if the driver is violating the road signs. This number plate is developed through the combination of embedded systems devices. It is composed of two parts, a SNP that can be fixed in front of the vehicle and the second part of a flexible display that can be set into the driver's cabinet near the vehicle dashboard. The first part smart plate contains the unique numbers that differentiate it from others, sensors including a speed sensor and Mini Smart Camera

A. Niyomugaba (✉) · N. Mduma · K. Michael
School of Computational and Communication Sciences and Engineering, The Nelson Mandela African Institution of Science and Technology (NM-AIST), Arusha, Tanzania

J. Marx Gómez et al. (eds.), *Artificial Intelligence Tools and Applications in Embedded and Mobile Systems*, Progress in IS,
https://doi.org/10.1007/978-3-031-56576-2_21

(MSC) that captures the road signs, Raspberry Pi 3B+ that acts as a microcontroller board to control these sensors, and processes the image of traffic signs and recognize them, and Mini Photo Voltaic cells (MPV) of solar panel as a source of current that charges its battery. Also, this unit contains a Global System Position (GPS) receiver for locating the vehicle and a Global System Mobile communication (GSM) for sending the notification through messages. And the second part display contains the screen that displays all information to the driver and the Buzzer that alerts the driver when required. The smart plate and its display intercommunicated through wireless communication. Based on the fundamentals of ANN technologies used to develop this SNP, the combination of the smart plate and display makes this SNP to be applied in traffic road signs monitoring since it can quickly identify and monitor the vehicle's location, vehicle's speed, road signs respected or violated by the driver, and detects an accident once it happens.

2 Related Works

The developed solution is based on the limitations of the previous related works made by other researchers. Various numbers of researchers provided many solutions for smart transportation, such as intelligent systems for assisting drivers and increasing road safety, Internet of Things (IoT)-based traffic systems by vehicles' number plate identification with traffic monitoring [5], and intelligent systems for vehicles number plate detection and recognition based on Convolutional Neural Networks [4]. However, the developed solutions were based on intelligent cameras that can read and recognize the vehicle's number plate from the front side.

Another similar study developed a solution for vehicle number plate detection using the improved algorithm [6]. The work focused on detecting Tanzania's plate numbers using the algorithm. The study provided the solution of localizing the plate area, extracting and segmenting the plate number's characters, and recognizing the vehicle's number plate. The MATLAB R201b simulation software was used to process the Hue Saturation Value (HSV) color space image, morphological matching, and statistical analysis operations to recognize the vehicle image and compute the plate number area. However, the solutions did not explain how the system should recognize the vehicle once it reaches where these intelligent cameras are not fixed.

Furthermore, other researchers proposed a vehicle's number plate recognition system with a web-based to recognize fast cars. The purpose of this solution was to prevent vehicles from violating traffic rules, assist traffic law enforcement agencies, improve road safety, improve road security, and reduce road accidents [7]. The developed system contains an infrared (IR) sensor that detects the vehicle's movement. If the distance is equal to the distance set, the camera triggered captures the number plate and stores the image on the Raspberry Pi. This image stored in Raspberry Pi gets processed within the system for character segmentation and recognition. However, the proposed solution did not focus on assisting the low traffic enforcement from the starting point up to the vehicle's destination.

An intelligent driver assistance system was also proposed with the ontology modeling approach for assisting vehicle drivers with a real-time warning message during critical situations. The system focused on generating alert messages based on context-aware parameters like driving situations, vehicle dynamics, driver activity, and environment [8]. However, the system did not assist the vehicle's drivers with road sign recognition. The study conducted at the Silesian University of Technology developed a mobile system that can detect and recognize traffic signs based on template matching [9]; the study consisted of two main stages like detection stage for detecting the road signs and the recognition stage for matching the detected signs with signs already stored into the database. However, although the system provided a solution for assisting drivers while driving, it did not accommodate functionalities without network connectivity.

3 Material and Methods

This section explains the system design, development, and communication protocols. This approach relies not only on communication architecture but also on the system architecture of the SNP is also introduced.

3.1 System Development Approach

The Agile Extreme Programming (XP) [10] system development methodology was adopted to develop this SNP. The Agile Extreme programming (Agile XP) is the framework selected from Agile System Methodologies since this framework allows a small team or individual to work on a project independently. This methodology contains five stages of planning, designing, programming, testing, and product increment for acceptance tests. Although Agile XP is straightforward to adopt changes within the project development phases, Agile XP has the advantage of testing the project from the initial stage by collecting the feedback that progresses the system's output. This model provides the competitive benefits of Agile XP of implementing a project that fulfills the customers' needs. The Agile XP also operates under five central values: communication, simplicity, feedback, courage, and respect.

3.2 *Artificial Neural Network System Development Approach*

Figure 1 describes the ANN computed system that can operate as human brains where the TensorFlow Model of machine learning and Python 3 were used to develop the system. TensorFlow model was used to process, classify, and recognize different objects, while Python programming language was used to program and compile the developed system. Therefore, with the TensorFlow model and Python 3, the real-time traffic signs recognition and notification system were developed. Based on ANN development approaches, the system takes the input data detected by sensors, processes, and generates the output to be notified to the dedicated user (driver or police station).

3.3 *The Algorithm Workflow of Artificial Neural Network-Based Smart Number Plate*

The system was developed based on three main inputs, speed detection, speed signs tracking, and vehicle location detection. Suppose the Speed sign detected indicates a lower than the current vehicle speed. In that case, the system will notify the driver if the speed remains, and the system will report the traffic station. Figure 2 presents the algorithm flowcharts of the Artificial Neural Network-based Smart Number Plate (ANN-SNP).

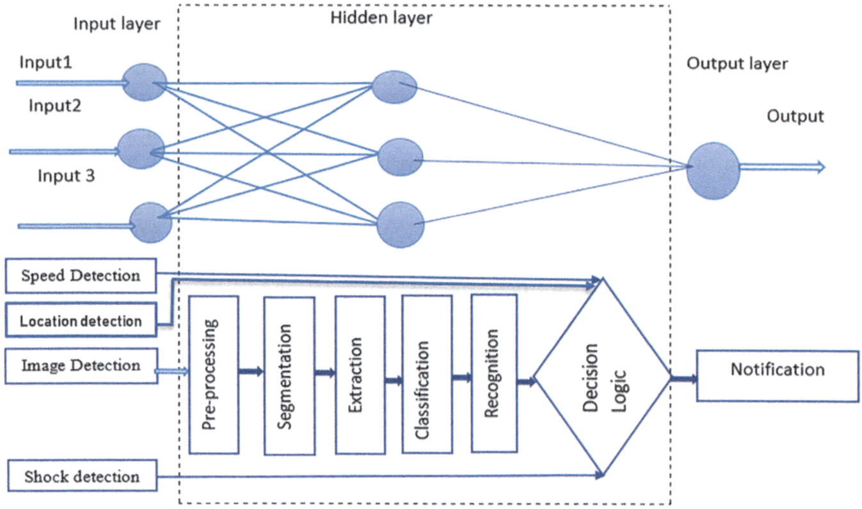

Fig. 1 Smart number plate's system workflow

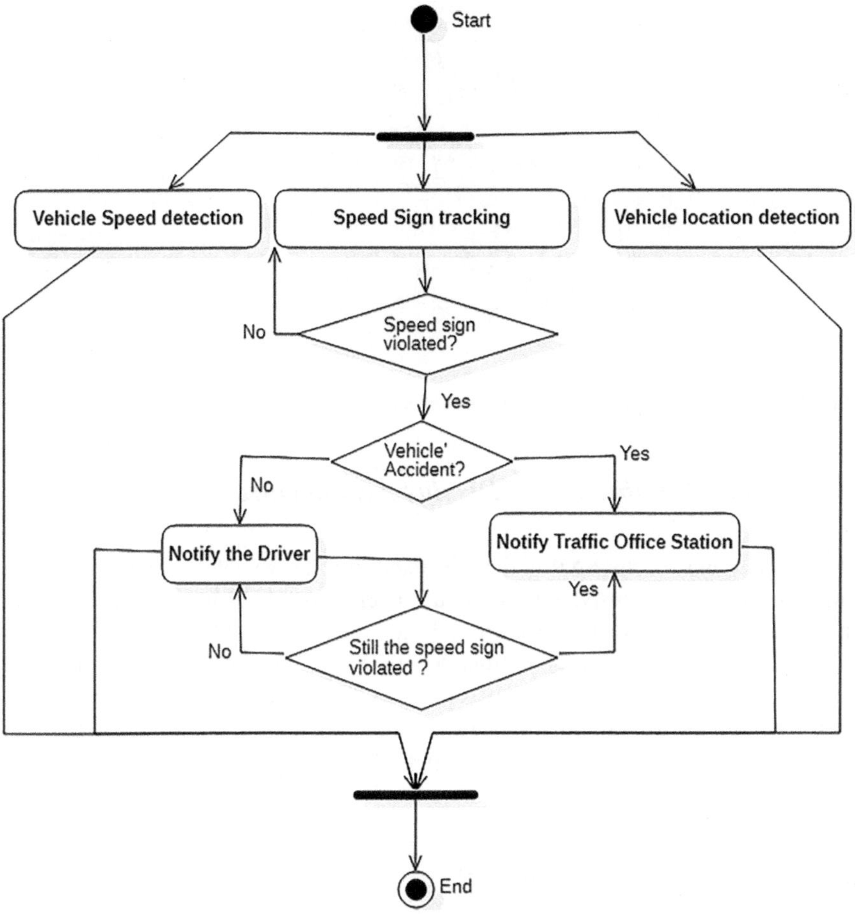

Fig. 2 Flowchart diagram of smart number plate

3.4 Smart Number Plate Architecture

The SNP was initially designed based on two primary principles of system architectures. First, the physical architecture that shows the interconnection of sensors and actuators to the microcontroller board and the logical architecture that shows the logical step-by-step functionalities of the system were designed. The physical architecture diagram was designed based on the combination of two units: detecting and processing units. The detecting unit comprises an Lm393 speed sensor, a global position system (GPS) NEO6M, and a vibration shock sensor. Raspberry Pi 3B+ and ESP32 DEVKIT make the processing unit. Figure 3 illustrates the schematical architecture of the SNP.

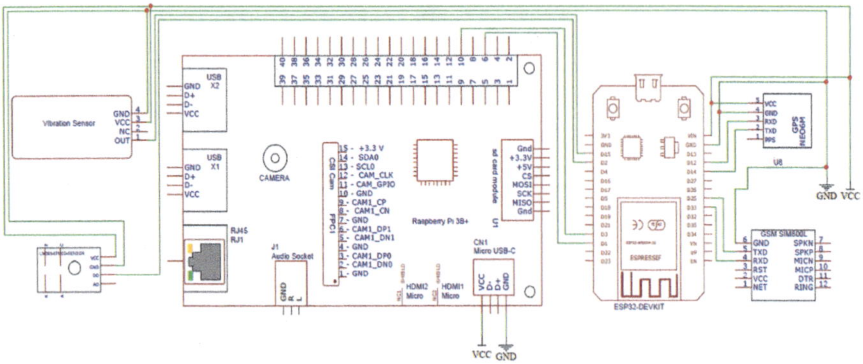

Fig. 3 The smart number plate schematical diagram

3.5 Smart Number Plate' Display Architecture

The display unit of the SNP is flexibly fixed into the vehicle's cabinet to display all information processed by the smart license plate. This unit combines the notifications devices such as Buzzer and Liquid Crystal Display (LCD). Therefore, to increase the efficiency of notification of SNP, the global system for mobile communication (GSM) SIM800L was interfaced with the SNP board as a third-party notification to notify and alert through messages. Figure 4 shows the schematical architecture of Smart Number Plate Display (SNPD).

3.6 Communication Protocol

The ESP-Now is a wireless communication protocol implemented by expressif technologies. It allows one-way and two-way communications between two or more ESP32 microcontrollers without Wi-Fi with packet transmission up to 250 bytes. In this case, the ESP32-S Board can simultaneously act as both the master (sender) and the slave (receiver). With ESP-Now, the pairing between microcontrollers is needed prior to their communication; after the pairing is done, the connection is safe and peer-to-peer with no handshake. Furthermore, this protocol operates with 2.4 GHz, making it a low communication power consumption [11]. This paper used the ESP-Now communication protocol to enable wireless communication between the SNP and its display fixed in the cabinet near the vehicle's dashboard. Figure 5 presents the ESP-Now wireless communication between two ESP32 microcontroller boards.

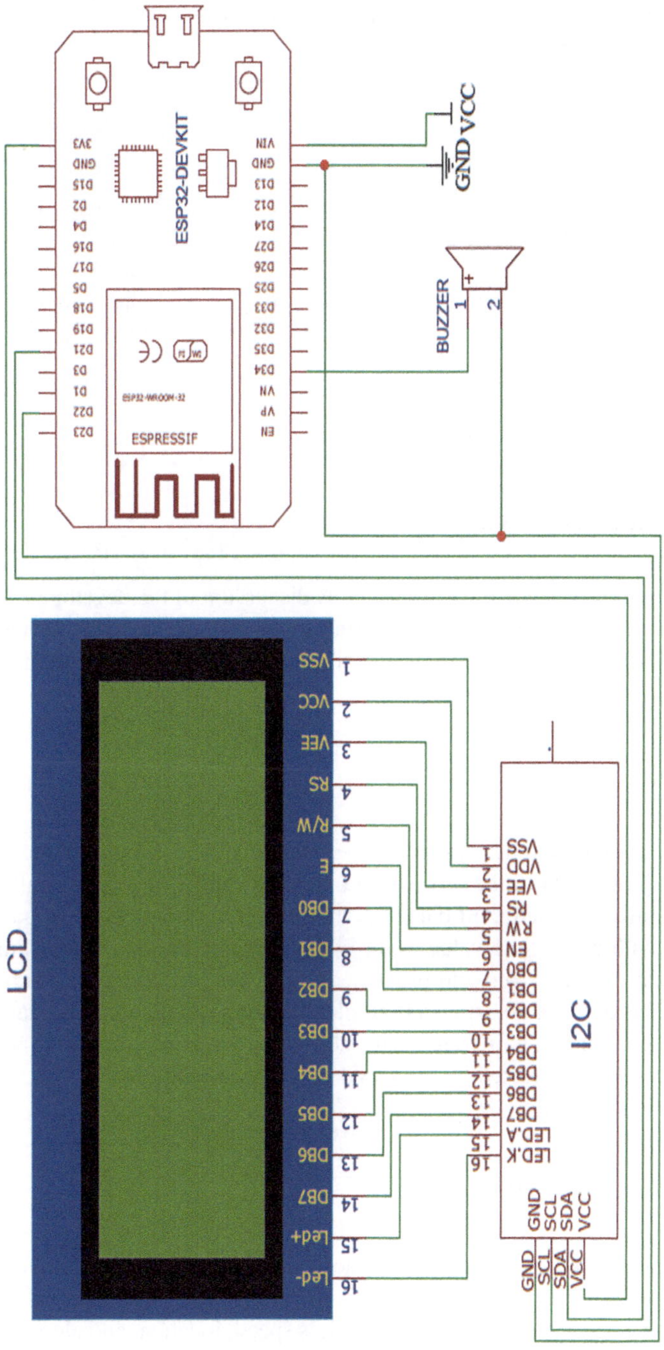

Fig. 4 The display of the smart number plate schematical diagram

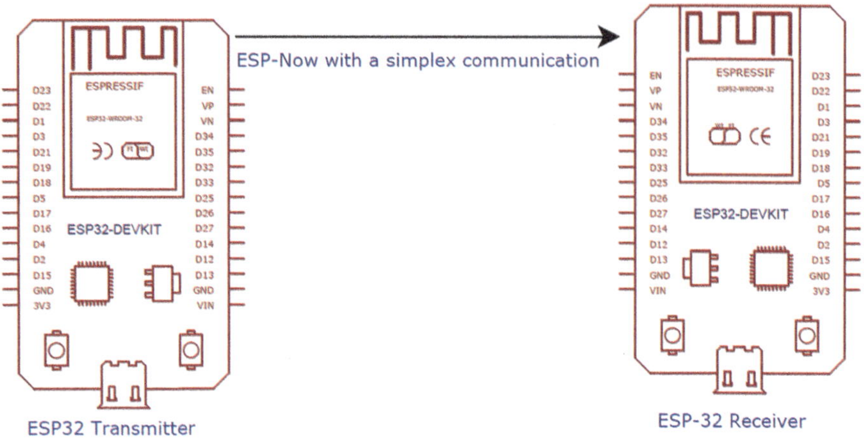

Fig. 5 ESP-Now wireless communication protocol

4 Results and Discussion

This section briefly presents the results and its discussion of the developed SNP per classified system units. Methods and materials involved have led to the development of an ANN-SNP that can recognize traffic signs with real-time notification.

4.1 Sensing Unit

The sensing operation unit was properly interfaced and produced the desired outcomes of digital data to the processing unit. The vehicle speed and traffic signs were successfully detected, and data was transmitted to the processing unit. Data were plotted with a real-time value using Python. Therefore, the data generated by the sensing unit was received and processed. Since the processing unit acts as the SNP's brain, tasks of data segmentation, classification, prediction, and generating output to the notification unit were performed. Figure 6 demonstrates the variation in the vehicle's speed detected by the speed sensor over time.

4.2 Notification Unit

The notification unit was properly interfaced and could notify the driver of the desired outcomes. In addition, the notification unit was able to send a Short Message Service (SMS) to display the vehicle's speed, location, and alert the driver against the violation of traffic signs while driving. Figure 7 shows a SNP display

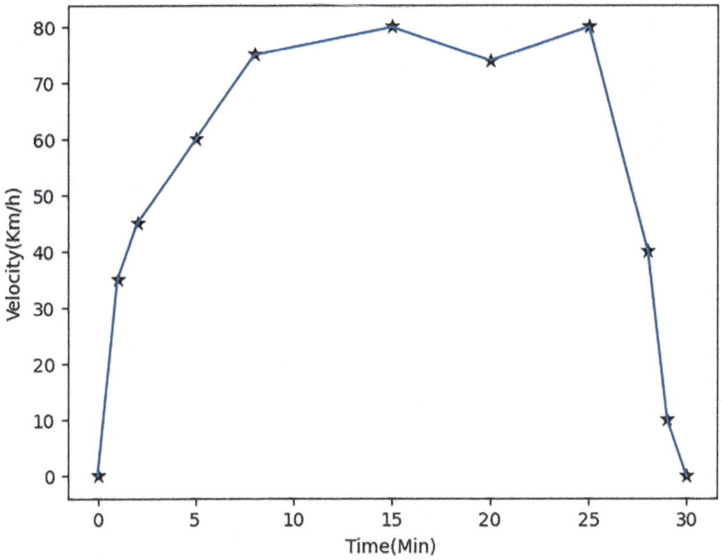

Fig. 6 The variation of vehicle's speed detected by the speed sensor

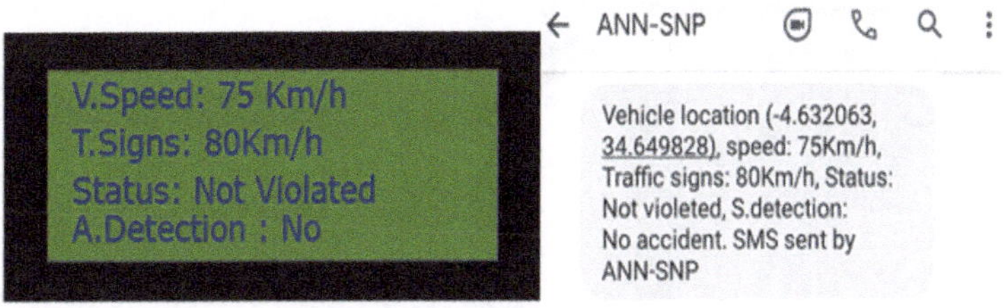

Fig. 7 The display and SMS from the Smart Number Plate (SNP)

and short message service (SMS) sent by the Global System for Mobile (GSM) communication.

4.3 Smart Number Plate (SNP)

Based on agile extreme programming (XP) development methodology, the testing and product increment were done, and the SNP produced the expected outcomes. In addition, the data transmitted was encrypted using MD5 encryption, and the hardware was secured in such a way if the SNP is removed from the vehicle by an unauthorized person, it immediately sends the notification and coordinates. Figure 8 shows the SNP.

Fig. 8 Smart number plate

ACCURACY

90.87%

	SPEED SIGN 60 KM/H	UNKNOWN	UNKOWN	UNCERTAIN
SPEED SIGN 60 KM/H	100%	0%	0%	0%
UNKNOWN	0%	100%	0%	0%
UNKOWN	0%	0%	88.3%	11.7%
F1 SCORE	1.00	1.00	0.94	

Feature explorer

Fig. 9 The system model testing and validation results

4.4 System Model Testing and Validation

The edge impulse was used to test and validate the system. The edge impulse is a TensorFlow compiler-based platform used to build, train, test, and validate various Machine Learning (ML) models. The edge impulse corrects data from raspberry Pi or other Mbed devices for data acquisition. The MobileNetV1 96×96 0.25 model was used to train and validate this model. MobileNetV1 96×96 0.25 is one of the convolutional Neural Network Networks designed for Embedded vision applications. Figure 9 presents the output results of the tested and validated model.

5 Conclusion

The missing road traffic signs in the transportation of goods and passengers may cause a dangerous situation, including the death of people and the destruction of infrastructures. Thus, the use of technology in traffic signs monitoring can play a vital role in increasing awareness of road safety and protecting the community against accidents. Therefore, smart traffic sign detection and recognition will enhance confidentiality and decrease the number of accidents in smart transportation. This SNP assist not only drivers and traffic police but also insurance companies and contributes to the country's economic growth. Moreover, the developed ANN-SNP is portable, can recognize the road speed limit signs and replace the traditional number plate for vehicles, especially those travelling a long distance, such as Buses for public transportation and tracks that carry wood. Other machine learning models for recognizing the remaining road signs, such as informative road signs, should be developed and added to the ANN-SNP system for future work.

References

1. Sagu, A., Gill, N. S., & Gulia, P. (2020). Artificial neural network for the internet of things security. *International Journal of Engineering Trends and Technology, 68*(11), 137–144. https://doi.org/10.14445/22315381/IJETT-V68I11P218
2. Ramasamy, L. K., & Kadry, S. (May, 2021). Internet of things (IoT). In *Blockchain in the Industrial Internet of Things*. https://doi.org/10.1088/978-0-7503-3663-5ch1
3. Rajput, S. K., et al. (2022). Automatic vehicle identification and classification model using the YOLOv3 algorithm for a toll management system. *Sustainability, 14*(15), 9163. https://doi.org/10.3390/su14159163
4. Alam, N. A., Ahsan, M., Based, M. A., & Haider, J. (2021). Intelligent system for vehicles number plate detection and recognition using convolutional neural networks. *Technologies (Basel), 9*(1), 9. https://doi.org/10.3390/technologies9010009
5. Patil, S., & Jadhav, M. (2019). IOT based traffic system by vehicle number plate identification & traffic monitoring. *International Journal of Scientific & Engineering Research, 10*(7), 309–311.
6. Nerey, E., Mvungi, N. H., Munuo, C. H., & Kisangiri, M. (2014). Vehicle Plate Number Detection and Recognition Using Improved Algorithm. *Computer Engineering and Intelligent Systems, 5*, 31–41. http://iiste.org/Journals/index.php/CEIS/article/view/16354
7. Shobayo, O., Olajube, A., Ohere, N., Odusami, M., & Okoyeigbo, O. (2020). Development of smart plate number recognition system for fast cars with web application. *Applied Computational Intelligence and Soft Computing, 2020*, 1–7. https://doi.org/10.1155/2020/8535861
8. Popova, I., Danilov, I., & Abdulina, E. (2020). Intelligent driver assistance systems as factor of transportation safety assurance. *Transportation Research Procedia, 50*(2019), 552–558. https://doi.org/10.1016/j.trpro.2020.10.066
9. Maćkowski, M., Sawiski, M., & Walczyszyn, W. (2019). Mobile system for road sign detection and recognition with template matching. *MATEC Web of Conferences, 252*, 03014. https://doi.org/10.1051/matecconf/201925203014
10. Shen, B. (2013). Agile methods (Scrum, XP) applying into small (micro) enterprise business website development. Computer and Systems Science (Information Systems) Media and Informatic Department Economic, p. 58.
11. Pasic, R., Kuzmanov, I., & Atanasovski, K. (2021). ESP-NOW communication protocol with ESP32. *Journal of Universal Excellence, 6*(1), 53–60. https://doi.org/10.37886/ip.2021.019

Intermodal Matching Algorithm Including Public Transportation and Ride-Hailing

Sven von Höveling, Cedrik Theesen, Anton Butenko, and Venkatesh Sasapu

1 Introduction

Statistics show that in Germany more than one in four is 60 years old or older. As a result of demographic change, this share will increase to more than one-third over the next few decades [1].

At a certain age, older people can no longer drive by themselves. Nevertheless, the need for mobility remains. What remains is public transport or individual transport services such as cabs or relatives. Rural areas, where more than 57 million people live in Germany, suffer from a lack of available bus connections [2]. In this chapter, we want to address the problem of mobility of elderly people. Our solution uses public transport and ride-hailing. The challenge is to combine these two transportation options.

Ride-hailing as part of ride-sharing is also being discussed and tested a lot but has not yet become widespread. The motivation behind this are usually sustainability and cost reduction.

This is important since the long-term mobility trend shows that people in Germany are spending increasingly more time commuting [2]. As a result of the increase in mobility, conurbations are facing the problem of traffic congestion. Particularly in heavily populated cities, this can lead to an increased air pollution. Diseases such as chronic obstructive pulmonary disease or asthma can be caused by it [3]. Globally, transport-related emissions were increased in the last decades. In

S. von Höveling (✉) · C. Theesen · A. Butenko · V. Sasapu
Carl von Ossietzky University of Oldenburg, Business Information Systems/VLBA, Oldenburg, Deutschland
e-mail: sven.von.hoeveling@uol.de; sven.von.hoeveling@uni-oldenburg.de; cedrik.theesen@uol.de; anton.butenko@uol.de; sasapu.venkatesh@uol.de

J. Marx Gómez et al. (eds.), *Artificial Intelligence Tools and Applications in Embedded and Mobile Systems*, Progress in IS,
https://doi.org/10.1007/978-3-031-56576-2_22

2015, the emissions due to transportation were 24% of total greenhouse gas released [4].

The work is written within the REMOBIAS project.[1] The goal of the project is to make mobility easier for elderly people by allowing them to easily plan routes by voice-based assist systems under the term smart living to meet doctors, relatives, shopping facilities, and other places themselves. For this purpose, a solution with integrated public transport, carpooling, and other types is considered for the target group of elderly people.

A student project group at the University of Oldenburg has developed a matching algorithm (MA) [5]. This algorithm is designed to allow people to join a suitable trip of somebody else as a passenger and was initially planned for the Instaride[2] project. The routing itself is realized via Openrouteservice[3] (ORS). If a trip is planned, it is possible to specify multiple filter options. These include, for example, social preferences. Different routes are then selected, evaluated, and compared on the basis of fixed parameters. This reduces the number of possible trips. A further reduction is achieved by isochrone matching. Isochrones are polygons for which points there is a limit of traveling time to reach them from the center.

For the REMOBIAS project, we now plan to modify the existing and well-working algorithm, so that intermodal trips targeted for older people are possible.

The work is structured as follows. In Sect. 2, we give a short overview of the challenges for intermodal trips, of existing approaches, and of similar projects, especially regarding our target group. Our concept is explained in Sect. 3 by first summarizing MA and then showing the planned modifications. Finally, in Sect. 4, a conclusion and an outlook follow.

2 Related Work

In the following, we will address the challenges and already existing approaches.

2.1 Challenges

Developing modal mobility chains remains a major challenge. If different mobility services are to be connected, it is also necessary to be able to communicate with several providers. From a technical point of view, there is not a de facto standard for interfaces that could be used. Another factor is that if only one provider changes its specification, the entire chain no longer functions.

[1] https://www.remobias.de

[2] https://www.instaride.de

[3] https://openrouteservice.org

Another problem is time-critical route planning. In case of disruption on one section (a train is delayed), all subsequent sections are also at risk and, in the worst case, have to be rescheduled: A chain is only as strong as its weakest link. Particularly in the case of carpooling, it is then often no longer possible to organize a suitable alternative. Some of these factors are known as the dial-a-ride problem [6].

In contrast to public transport, carpooling poses completely different problems. There are no timed journeys. Individual driving styles or the inclusion of additional passengers leads to an unpredictable journey time. In addition, trips must be arranged in advance. You cannot just hop on like a train. Taking these special features into account, automated multimodal route planning is a major challenge. Rating systems have asserted themselves in the case of carpooling. Considering these rating systems makes the automated planning process more difficult. However, surveys from the NEMo[4] (sustainable fulfillment of mobility demands in rural areas) have shown that trust in the driver is a high priority [7].

The challenge can also be reasoned theoretically as the intermodal case adds some complexity to the typical ride-matching problem that is already NP-hard as it can be subordinated to the dial-a-ride problem (DARP) [5, 8].

2.2 Approaches

With the NEMo[5] project, an attempt to turn citizens in rural areas into mobility providers themselves in order to compensate for the poor bus infrastructure has already been made with the Fahrkreis app. One of the results was that a critical mass of users must be built up for such a system to work at all. NEMo, as a research project, was not in a position to take measures (advertising) to build up such a mass. Fahrkreis has the status of a prototype.

Back in 2013, Deutsche Bahn launched the multimodal mobility platform Qixxit. By default, all means of transport from airplane to cab were included in the route determination. If you generally do not want to use certain means of transport, you can deselect individual ones. In 2019, the platform was sold to LM Group and ceased in 2021. The reason was not specified and there was too little content in [9].

[4] https://nemo-mobilitaet.de

[5] https://nemo-mobilitaet.de

3 Simple Intermodal Matching Algorithm (SIMA)

In this section, we show the algorithm SIMA. First the base algorithm MA of [5] is shown. Then its usage in an intermodal case is explained, followed by some ideas to reduce the execution time.

3.1 *Matching Algorithm*

Assume that there is a list of trips some drivers planned, typically a shortest trip from one point to another. Moreover, assume that those drivers agree to take people (in the following passengers) with them and to drive a certain detour. With MA it is now possible to find trips for passengers to join (matchings), possibly changing the trip of a driver. A passenger needs to provide a request consisting of a time interval, a pick-up, and a drop-off point. The results are ranked based on a metric taking into account the detour for the driver and some social parameters. Both the driver and the passenger can filter the possible matching by filters like the minimal age or the gender of a passenger/driver.

One of the core ideas of the algorithm is to reduce the number of possible matching using isochrones. Assume the time interval fits, and there are many potential tours to join for the passenger. Recalculating a trip using a trip planner like Openrouteservice[6] is relatively expensive in terms of time. Hence, before a recalculating is done, the number of potential trips needs to be reduced. For this, isochrones of the pick-up and drop-off points are used. Isochrones of different time levels (up to 10 minutes) are calculated, and if a tour does not intersect with these, a trip can be removed from the potential ones. This reduces the number of possible trips. In Fig. 1, the idea of the isochrones and its usage to filter are visualized.

Other important steps during the filtering are a check of the order of the isochrones intersections and a check of spatial limits. The first ensures that drop-off point is not before the pick-up point on the route of the driver. The latter ensures that the driver is not starting his own trip into opposite direction to pick somebody up and that he is not passing his own end point to drop somebody off. More details can be found in the paper of the student group for that algorithm [5]. See Algorithm 1 for an outline.

3.2 *Enabling Simple Intermodal Matchings*

We now describe how we plan to use MA for an intermodal case. At the beginning, this is only planned for a certain model region, probably the region of Oldenburg, which limits the number of possible calculations and pre-processing. Assume a passenger wants to get from a pick-up point to a drop-off point, and there is no

[6] https://openrouteservice.org

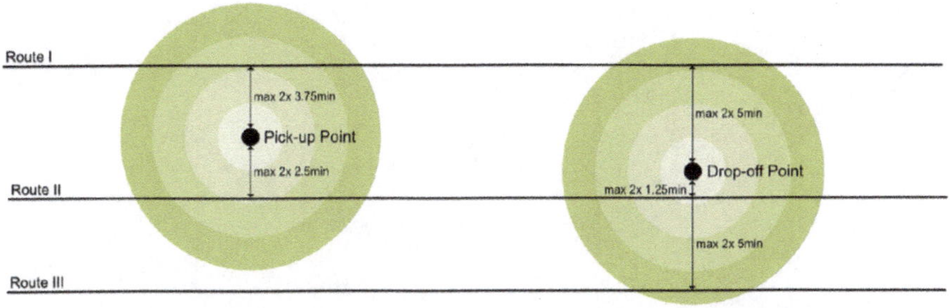

Fig. 1 A passenger wants to join a trip. The pick-up and drop-off points are known. For these points, isochrones are calculated, shown as green circles. The innermost circle corresponds to a maximum detour of 2.5 minutes (two times 1.25 minutes). The others correspond to 5, 7.5, and 10 minutes. Note that circles are used for simplicity, and in fact polygons are calculated with respect to the travel time along edges. For the available three routes, it is then checked whether they intersect with the circles, respectively, with the polygons. Recalculating the route I for carrying the passenger would mean detour upper bounded to 17.5 minutes. For route II, the upper bound is 7.5 minutes. Route III is not considered further since it is not intersecting with any isochrones of the pick-up point. Note that for route I and maybe even for the route III the actual detour calculated by a trip planner might be less than for route II. The figure is taken from [5]

Algorithm 1 Outline of MA based on the description and illustration of [5]

Require: Personal settings D and P of driver and passenger
Require: Request R of passenger
Require: List L of potential trips to join

```
Filter L by timeframe of R
Filter L by personal filters of D and P
Filter L by spatial limits
Calculate isochrones I for pick-up and drop-off points of R
Remove trips of L that do not match the isochrones I
Check order of isochrones intersections
for trip l of L do
    Calculate new route for the driver considering pick-up and drop-off points of R
    Calculate route dependent score for l
    Calculate social score for l
end for
Sort L depending on the scores and return L
```

---

suitable option by public transportation only and by ride-hailing only. Now, it might be possible to combine both. For simplicity and because for first acceptance tests it is enough, we limit the number of changes between both transport options to one. This results in exactly two cases: (i) The passenger first uses ride-hailing from the pick-up point and then uses public transportation to get to the drop-off point. (ii) The passenger first uses public transportation to start from the pick-up point and then uses ride-hailing to get to the drop-off point. While it is quite easy to check whether public transportation is available for (i) or (ii), it is more complicated to check whether available trips by drivers fit for the other part. If the intermediate point, the point where the passenger changes, is known, MA can be used. However,

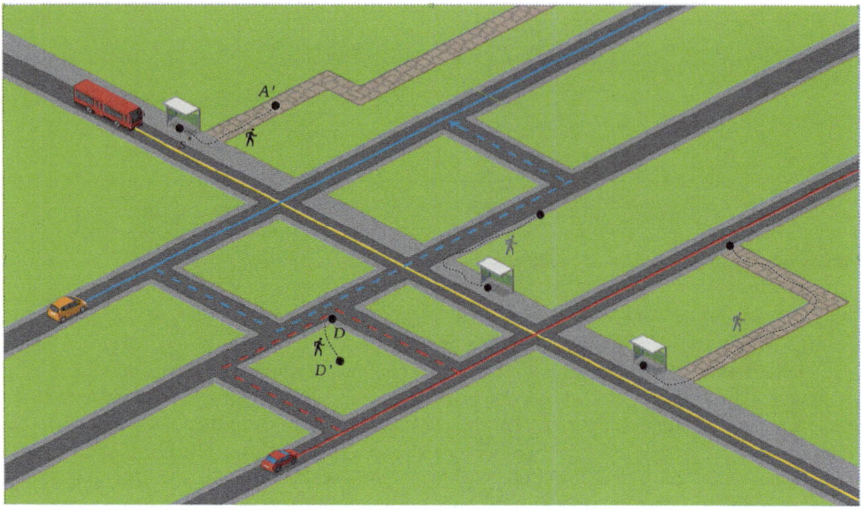

Fig. 2 Intermodal matching representation for the ride-hailing first case. 1. From the initial location D' passenger walks to the pick-up point D (dashed black line) that is located in vicinity from two car routes (solid blue and red lines). 2. Drivers change their routes (dashed blue and red lines) in order to take the passenger and bring him to the drop-off points. 3. Passenger walks to the bus stop. 4. Passenger travels by bus (solid yellow line) to the target bus stop s^*. 5. Passenger walks to his final destination A'

since this point is not known, the idea is to use different station points of the public transportation. Figure 2 visualizes this.

The main idea of intermodal matching is coupling two ways of transportation by using public transport stops as initial locations or final destinations for MA.

The base matching algorithm works with the concept of meeting points that are preselected locations for pick-up and drop-off. These points are chosen in order to make getting in and out of the car easier and faster. Good examples are gas stations and parking spaces. Also street intersections are possible. Required factors are, among others, parking opportunities and sufficient illumination. Finding the nearest meeting point leads to the nearest neighbor search problem that can be solved by using Voronoi diagrams. Within pre-processing, meeting points are determined as well as Voronoi cells around them. Thus, there will be a walking path during the transfer ride-sharing/public transport. However, if it is necessary for elderly people, walking can be avoided by adding all the stops into the set of meeting points.

In the following, we describe the intermodal matching problem and a solution concept in a more formal way. If $S = \{s_1, s_2, \ldots, s_n\}$ are the public transport stops, then for each stop and each moment of time $T : (s_i, s_j, d_k) \rightarrow a_{ijk}$, where s_i is a departure stop, s_j is an arrival stop, d_k is a departure time, and a_{ijk} is an arrival time. Relation T can be obtained from the transport timetable or the corresponding API that is provided by many transport organizations nowadays. In both cases (i) and (ii), it is necessary to determine the public transport stop nearest to the user's final destination and initial location, respectively. There are two possible approaches here. The first approach is the construction of planar Voronoi diagram with stops as the seeds. In this case, only one nearest stop will be considered in further algorithm steps. For the second approach, all stops reachable within certain walking time from the given point are considered.

The standard matching algorithm finds a list of potential routes $\{R_i\}$ for a given tuple (D, A, t, u) representing locations of departure and arrival, time of departure, and user preferences. This list is sorted by some criterion, normally travel time $a(R_i)$, and truncated to several options selected by the user. Let us consider the two above-mentioned cases, while the passenger wants to travel from location D' to A', which are the mentioned nearest stops of D and A. For simplicity, we consider only one transfer and one nearest public transport stop where the passenger walks to/from. In the following, an outline of the concept is presented. For simplicity, the topics social filters and preferences for (older) people are excluded. They will be covered in further research:

1. Ride-hailing first:

    1.1 Determine the nearest pick-up point D from D' with the Voronoi diagram.
    1.2 Find $s^* \in S$—stop that is nearest to A' and $s = \{s_i\} \subseteq S$—stops that have direct connection with s^* (without transfer).
    1.3 With the time step $\delta : d_k = d_{k-1} + \delta$, find R_{ik} and $a(R_{ik})$ from $(D, \mathbf{s}, d_k, u)$ with matching algorithm.
    1.4 Compute total travel time $a_{ik} = a(R_{ik}) + T(s_i, s^*, d_k)$. Let p be a number of maximum travel options shown to the user with the travel times $a_1, ..., a_p$. Then $a^* = \max_{m\in(1,p)} a_m$ will be used as threshold to decide whether to continue search along the time axis for given s_i.
    1.5 Repeat the steps 1.3, 1.4 until all $a(R_{ik}) \geq a^*$ or $a_{ik} \geq a^*$.

2. Public transport first:

    2.1 Determine the nearest drop-off point A from A' with the Voronoi diagram.
    2.2 Find $s^* \in S$—stop that is nearest to D' and $s = \{s_j\} \subseteq S$—stops that have direct connection with s^*.
    2.3 With algorithm, find R_{jk} and $a(R_{jk})$ from (s, A, d_k, u) for $d_k = d_{k-1} + \delta$.
    2.4 Compute the total travel time $a_{jk} = a(R_{jk}) + T(s^*, s_j, d_k)$; compute a^* analog to 1.4.
    2.5 Repeat the steps 2.3, 2.4 until all $a(R_{jk}) \geq a^*$ or $a_{jk} \geq a^*$.

Thus, SIMA is an iterative process of MA execution with different input parameters: spatial and temporal. Practically, the matching algorithm should be implemented or used in a way that it allows for multiple requests and a parallel computation.

An outline of the overall transportation algorithm of the REMOBIAS project can be found in the appendix.

3.3 *Possible Adjustments for Shorter Execution Time*

For MA, the student group reached execution times of less than five seconds in many test scenarios. SIMA uses a modified version of MA several times, so the question arises if the execution of SIMA will be still acceptable. Literature suggests us to not take longer than ten seconds to not lose the attention of the user [10]. This might be hard to reach; however, in our case, the intermodal options are only the third and last options, after public transportation and ride-hailing only, which may excuse for some more seconds of waiting. To reach better execution time, we have developed some strategies that could be further examined and implemented.

As the number of safe points is limited, isochrones can be pre-calculated. In the intermodal case, public transportation station points are either shifted to safe positions (meeting points) as well or they can be seen as safe points themselves. In either case, isochrones can be pre-calculated. Typically, Euclidean measurement is used for Voronoi diagrams. For better results in transportation context, alternatives such as in [11], which take into account the available streets, should be considered. Due to the target group and based on preferences, transportation options may be deleted if the shift toward a safe point is too large.

The highest impact probably offers the mentioned parallel computing, in particular of MA or the route calculation. For this, the related services can also be hosted multiple times.

Other ideas are more stop conditions within the algorithm, e.g., to stop further calculations if a certain quality feature is reached and pre-loading of public transportation tables. Regarding the public transportation stations, it is conceivable to lower their number by, e.g., considering every second station.

If the calculation of the intersections between existing routes and isochrones is time critically, the routes and isochrones could be smoothed before. Since an intersection of routes and public transportation needs to be found, more advanced spatial pre-calculation regarding their positions to each other is conceivable.

4 Conclusion and Outlook

In this chapter, we presented a concept for an intermodal algorithm combining public transportation connection and ride-hailing. The number of changes between both transportation options is limited to one, and the concept is thought to be used for a limited model region first and the target group of older people. The algorithm uses a modified version of the algorithm in [5] as subroutine that uses isochrones to achieve a shorter detour for the driver and route as well as social scoring. Hereby, one input point (pick-up or drop-off point) is used as start or end point for the public transportation.

Next, an implementation is planned. Depending on the resulting execution time, the usage of presented acceleration possibilities will be considered. The implementation is followed by an evaluation within the project REMOBIAS. The focus is on the general acceptance of the intermodal trips with respect to the stated target group, which reasons the rather simple concept of the algorithm.

Parts of the concept should be able to be generalized or transferred to other use cases.

Appendix

Algorithm 2 Outline of the overall algorithm of the project REMOBIAS

```
Require: Request R of passenger
Require: Personal settings D and P of driver and passenger
Require: List L of potential trips to join
Require: List P of public transportation options
if suitable (time, personal preferences) public transportation options available then
    Return options ranked by some metric, taking into account route and social aspects
end if
if suitable (time, personal preferences) ride-hailing options available (MA) then
    Return options ranked by some metric, taking into account route and social aspects
end if
if suitable (time, personal preferences) intermodals options available (SIMA, both cases) then
    Return options ranked by some metric, taking into account route and social aspects
else
    Return empty list of options
end if
```

References

1. Federal Statistical Office: Older people in Germany and the EU, 2016. https://www.bmfsfj.de/resource/blob/113952/83dbe067b083c7e8475309a88da89721/aeltere-menschen-in-deutschland-und-in-der-eu-englisch-data.pdf, Cited 13 Sep 2022.
2. Küpper, P. (2016). *Abgrenzung und Typisierung ländlicher Räume*. Johann Heinrich von Thünen-Institut.
3. Manisalidis, I., Stavropoulou, E., Stavropoulos, A., & Bezirtzoglou, E. (2020). Environmental and health impacts of air pollution: A review. *Frontiers in Public Health, 8*, 1–2.
4. Scheelhaase, J., Grimme, W., O'Sullivan, M., Naegler, T., Klötzke, M., Kugler, U., Scheier, B., & Standfuß, T. (2018). Klimaschutz im Verkehrssektor – aktuelle Beispiele aus der Verkehrsforschung. *Wirtschaftsdienst, 98*(9), 655–663.
5. Eilers, M., Hempen, S., Frers, S., Stahr, B., Hölscher, N. H., Bürmann, P., Janßen, C., & Marx Gómez, J. (2021). An instant matching algorithm in the context of ride-hailing applications, using isochrones and social scoring. In *INFORMATIK 2021*. Gesellschaft für Informatik (pp. 103–114). https://doi.org/10.18420/informatik2021-007.
6. Nils Woitzik. (2007). Das Dial-a-Ride-Problem. https://www.zaik.de/AFS/teachings/ws0708/ORSeminar/aus-arbeitungen/handout_DialARide_NilsWoitzik.pdf. Cited 15 Sep 2022.

7. Jahns, M., Woisetschläger, D. M., & Seegebarth, B. (2018). Never trust a stranger – Even if (s)he is beautiful? – How attractiveness, gender, and role of associates influence ridesharing intention. In *Proceedings of the European Marketing Academy Conference*, Glasgow.
8. Campbell, I., Ali, M. M., & Fienberg, M. L. (2016). Solving the dial-a-ride problem using agent-based simulation. *South African Journal of Industrial Engineering, 27*, 143–157.
9. Reise vor9. (2021). LM Group stellt Mobilitätsplattform Qixxit ein. https://www.reisevor9.de/marketing/lm-group-stellt-mobilitaetsplattform-qixxit-ein. Cited 13 Sep 2022.
10. Nielsen, J. (1994). *Usability engineering*. Morgan Kaufmann Publishers Inc.
11. Butenko A., & Marx Gómez, J. (2022). Combined algorithm for Voronoi diagram construction in application to dynamic ride sharing. In *MOBILITY 2022: The Twelfth International Conference on Mobile Services, Resources, and Users* (pp. 5–8), Porto, Portugal.

A Machine Learning-Based IoT Environmental Monitoring Platform for Data Centers

Wayne Steven Okello, Silas Mirau, Michael Kisangiri, Andrew Katumba, and Edwin Mugume

1 Introduction

Monitoring is always paramount in any organization irrespective of its structure and nature of its business [1]. Monitoring enables early identification of prospective problems so that they can be promptly addressed before causing severe damage and thereby ensuring reliable execution of company operations.

A data center refers to a facility that houses computing resources, networking, and storage infrastructure as well as associated utilities such as power supplies and Heating, Ventilation, and Air Conditioning (HVAC) systems used by business entities to assemble, process, store, share, and access vast quantities of data. Data centers vary in size from as small as an exclusive space within a building to as large as a group of buildings, depending on various factors that include but are not limited to company size and core business of the organization.

Data centers were initially owned by mostly top technology firms but due to the increasing digitization, the evolution of new technologies such as Cloud Computing, Artificial Intelligence (AI), Internet of Things (IoT), Big Data, and Blockchain, in addition to the significantly growing technology acceptance across the world today, business operations are more data driven than ever before and data centers have

W. S. Okello (✉) · S. Mirau · M. Kisangiri
School of Computational and Communication Science and Engineering (CoCSE), Nelson Mandela African Institution of Science and Technology (NM-AIST), Arusha, Tanzania
e-mail: okellow@nm-aist.ac.tz; silas.mirau@nm-aist.ac.tz; kisangiri.michael@nm-aist.ac.tz

A. Katumba · E. Mugume
Department of Electrical and Computer Engineering (ECE), Makerere University, Kampala, Uganda
e-mail: andrew.katumba@mak.ac.ug; edwin.mugume@mak.ac.ug

J. Marx Gómez et al. (eds.), *Artificial Intelligence Tools and Applications in Embedded and Mobile Systems*, Progress in IS,
https://doi.org/10.1007/978-3-031-56576-2_23

subsequently become a very crucial component in the business models of many modern enterprises [2].

There are a number of causes of unexpected data center outages that include but are not limited to human errors, power outages, environmental factors, network failures, and cyberattacks, with environmental issues contributing to 25% of downtime according to [3]. Environmental conditions such as extreme temperatures, presence of water, power spikes, and surges as well as fires in the data center can lead to equipment failure, damage, and loss of huge volumes of critical data. Low humidity leads to static discharge as a result of electrostatic energy accumulation which damages equipment and causes loss of data, while high humidity results in condensation as well as corrosion of electrical components.

2 Related Work

A number of researchers have tackled challenges in regards to environmental monitoring of data centers and some of them are highlighted in this section.

Kurniawan et al. [4] developed an IoT-based system that utilizes Wemos DHT Shield wireless sensor to capture the temperature and humidity readings. A Raspberry Pi processes the readings stored in a MySQL database and triggers the web server that enables the website with real-time graph visualizations to be accessible through the Internet or a private wireless network. Furthermore, the system uses the Python yowsup library to request data from the web server on the Raspberry Pi, processes it and finally returns the required information in the form of WhatsApp notifications.

Roy et al. [5] present a server room cooling system that uses the single bus wire protocol to connect a DS18B20 sensor to a Raspberry Pi MCU. The temperature is captured by the DS18B20 sensor which sends the sensed data to the Raspberry Pi, which compares it to database threshold value and dispatches a warning email notification message to the data center operator in case the temperature is below the threshold.

Onibonoje et al. [3] in their work propose a real-time conditioning and control system for server rooms based on IoT that monitors environmental factors namely heat, smoke, water leakage, power outage, light out, flame, and correlates the heat index with the area air temperature and humidity. It further undertakes timely corrective measures and sends out notifications to the administrator in case of any unexpected event.

Kaliyamurthie et al. [6] describe a wireless temperature and humidity measurement and monitoring system for data centers that utilizes LoRaWAN communication link for data transmission. The LoRaWAN gateway sends the sensed data to a cloud application for further processing and visualization, allowing more optimization of the air conditioning and ventilation systems.

Yumang et al. [7] describe a ZigBee-based system that employs thermal imaging to monitor server room temperature and humidity in real time. A microcontroller processes the data gathered by the infrared and humidity sensors which is sent to the computer via ZigBee. MATLAB software is then used to apply the color mapping and Bicubic algorithms and a buzzer alarm will be triggered in case the threshold value is reached.

Rasool et al. [1] present an RF-based centralized system that monitors temperature, humidity, battery life and status, water leakage, and smoke in server rooms. The daily data from the deployed sensors is transmitted using the Peripheral Interface Controller (PIC) to the Graphical User Interface (GUI) where it is stored and displayed to users in the form of tables and visualizations.

Khan et al. [8] propose a physical and environmental monitoring system based on IoT that comprises Raspberry Pis as smart objects and sensors located at various locations in the data centers to monitor temperature, humidity, and atmospheric pressure. The proposed system uses LoRa gateways to link the smart objects to the server and activates alarms upon detection of any undesirable situation.

Nasution et al. [9] highlight the design and development of a remote temperature and humidity monitoring system that uses a small supercomputer in the form of Lattepanda to which a DHT11 sensor is connected to read temperature and humidity values in the server room. The captured readings are sent to the cloud-based ThingSpeak platform every 30 s for storage and visualization.

Yamanoue [10] presents an IoT-based system comprising Raspberry Pi, sensors, Wiki pages, and bots that monitor servers and server rooms by utilizing a Network Address Translation (NAT) router to configure as well as manage sensors using a Wiki page.

From the related work, there are a number of environmental monitoring systems for data centers or server rooms. However, they have no prediction/forecasting capabilities, most of them only monitor temperature and humidity, offer no redundancy, and are limited in scalability. Therefore, there was a gap in developing scalable data center environmental monitoring systems with prediction capabilities in addition to keeping track of other environmental parameters other than temperature and humidity that can aid in the planning and optimization of data center resources.

3 System Implementation

The system comprises a heterogeneous WSN with four (4) sensor nodes and a sink node as illustrated in the system architecture in Fig. 1.

The system was further decomposed into five (5) components or units demonstrated in Fig. 2.

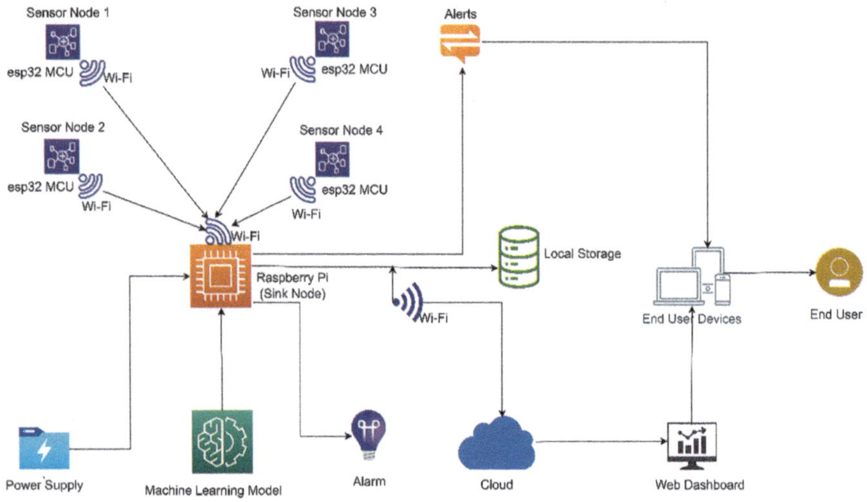

Fig. 1 System architecture describing conceptual interactions of the sensor nodes, sink node, and end user

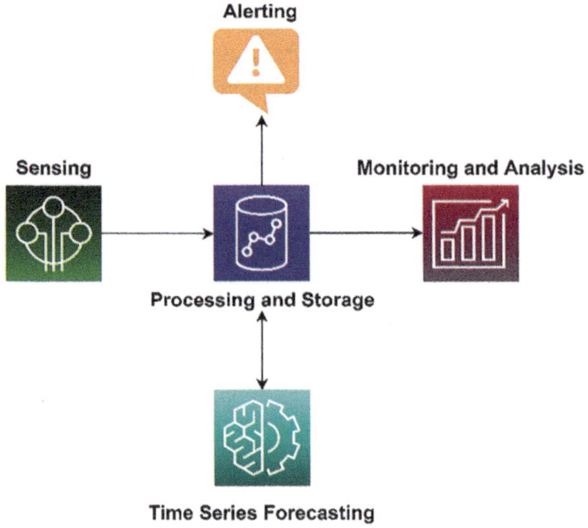

Fig. 2 Block diagram defining interrelations among system units

3.1 Sensing Unit

Four sensor nodes were set up and configured as follows.

Sensor Node 1 and Sensor Node 2

Sensor Node 1 and Sensor Node 2 are identical where the former is used for rack monitoring while the latter is used for ambient monitoring. They are composed of an ESP32 microcontroller, a DHT22 sensor, an MQ2 sensor, a 20x4 LCD display module, and a buzzer. The DHT22 measures temperature and humidity readings, the MQ2 sensor detects presence of smoke, the 20x4 LCD display module displays sensor readings and the buzzer produces an audio alarm in case (a) either temperature or humidity is outside the optimal data center range (temperature below 15 °C or above 32 °C and humidity below 20% or above 80%) [11], or (b) if smoke is detected.

Sensor Node 3

This sensor node consists of an ESP32 microcontroller, a water-level sensor for detecting the presence of water, a 20x4 LCD display module for displaying sensor readings and a buzzer to produce an audio alarm in case water is detected.

Sensor Node 4

It is composed of an ESP32 microcontroller, a ZMPT101B voltage sensor, an ACS721 current sensor module, a 20x4 LCD display module, and a buzzer. The ZMPT101B voltage sensor measures input AC voltage, the ACS721 current sensor module measures input AC, the 20x4 LCD display module displays sensor readings and the buzzer produces an audio alarm in case of voltage spikes (voltage exceeds 245 V). A multimeter was used in calibration of the ZMPT101B voltage sensor whose potentiometer was adjusted until sensor readings matched those of the multimeter.

3.2 Processing and Storage

A Raspberry Pi 4 was set up as the sink node. The sensor nodes send or publish their readings to a local MySQL database via Wi-Fi by making an HTTP POST request to the Raspberry Pi LAMP server. The sensor readings are inserted into the database every five (5) min and the database is backed up to the local SD card of the Raspberry Pi as well as Google Drive after every 24 h using the SqlBak service. The Raspberry Pi can be remotely accessed through Secure Shell (SSH), Virtual Network Computing (VNC), and TeamViewer. SSH and VNC can be used to access it while it is on the same network with the connecting device while TeamViewer can be used to access it from anywhere across the globe as long as both the Raspberry Pi and the connecting device have an Internet connection.

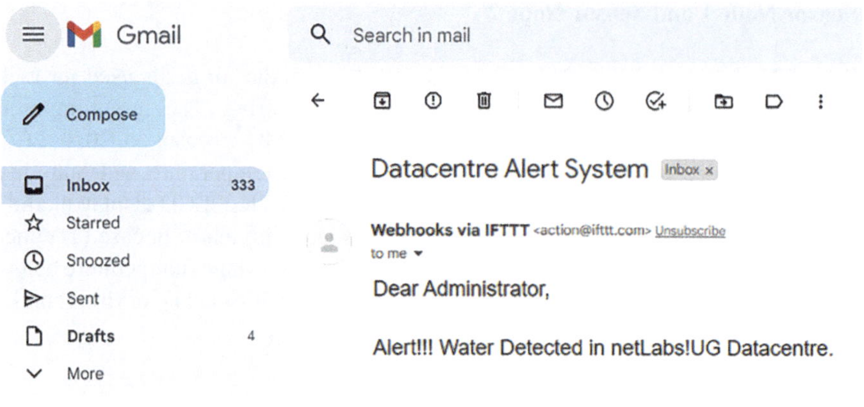

Fig. 3 eMail alert notification

3.3 Alerting

Each sensor node consists of a buzzer that produces an audio alarm in case an undesired event is detected. In addition, sensor nodes are integrated with email, SMS, and WhatsApp alerts that are sent to data center administrators in the event of unfavorable environmental conditions in the data center as illustrated in Fig. 3. The IFTTT web service was used in the email and SMS alert integration while Callmebot WhatsApp bot API was used for the WhatsApp alert integration.

3.4 Monitoring and Analysis

Sensor readings are sent to the ThingSpeak platform using the ThingSpeak API for cloud data logging and are updated every five (5) min. A customized web dashboard that utilizes the ThingSpeak platform iframes, developed using the dash framework, displays real-time trends of the environmental parameters measured by the different sensors as depicted in Fig. 4.

3.5 Time Series Forecasting

The dataset was collected from sensor node 1 deployed in the netLabs!UG data center, Kampala, Uganda where sensor readings were logged every five (5) min for a period of 44 days (from 11/07/2022 to 25/08/2022). The dataset's features are summarized in Table 1.

This phase involved data preprocessing, exploratory data analysis, modelling, and evaluation.

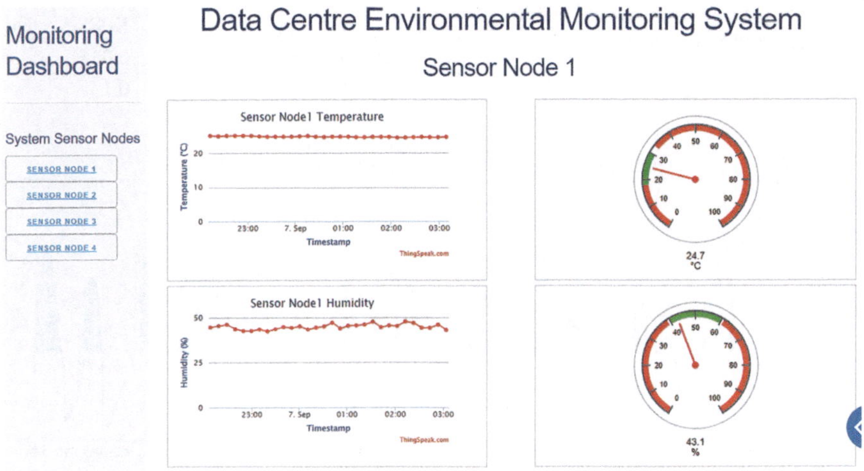

Fig. 4 Web dashboard displaying sensor node 1 temperature and humidity trends

Table 1 Dataset description

S. No.	Feature	No. of observations
1	Id	8010
2	date_time	8010
3	Temperature	7998
4	Humidity	7998

Data Preprocessing

This involved inspection of the panel data captured from the sensor nodes, data cleaning, data validation as well as data transformation. The aim was to get a better understanding of the data and to prepare it for training the machine learning models. The dataset was checked for missing values and it was discovered that there were twelve (12) missing values found. Linear interpolation was utilized to handle the missing values. Systematic outliers were observed in the humidity data where the humidity reading was always 25.5 °C whenever the temperature was 25.5 °C. These observations were dropped from the dataset.

Exploratory Data Analysis

Matplotlib and seaborn libraries were used to create visualizations that provided more insights about the panel data that in turn guided in the building of the machine learning models. The stationarity of the time series was checked using the Augmented Dickey–Fuller (ADF) test as well as Kwiatkowski-Phillips-Schmidt-Shin (KPSS) test where both tests confirmed that the series were stationary. The results of the tests are illustrated in Table 2 and Figs. 5 and 6.

Table 2 p-Values from the ADF and KPSS stationarity tests

Parameter	ADF p-value	KPSS p-value
Temperature	5.44e-05	0.055
Humidity	7.50e-08	0.1

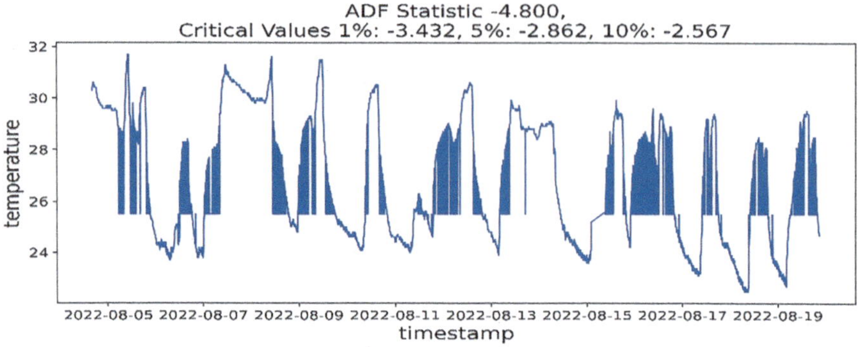

Fig. 5 Temperature distribution with ADF statistic and critical values

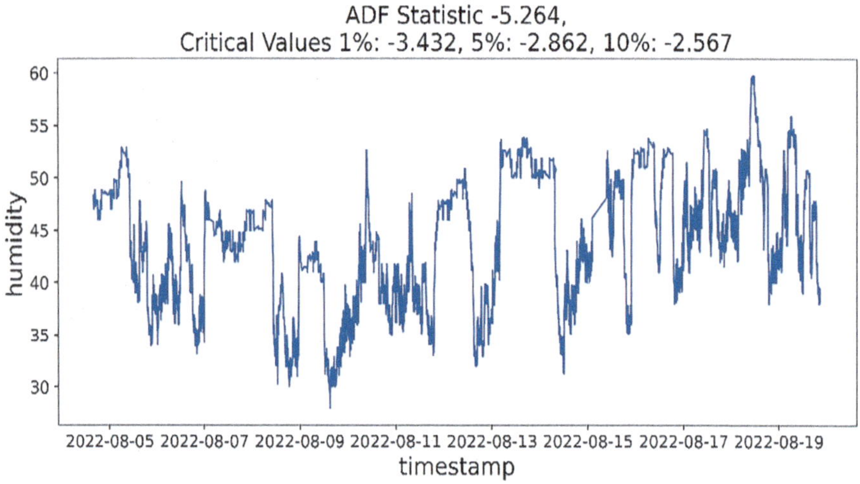

Fig. 6 Humidity distribution with ADF statistic and critical values

Modelling

The dataset was split into a train dataset and test dataset in a ratio of 0.8:0.2, respectively. Temperature and humidity were the target features. ARIMA, Facebook Prophet, and ES algorithms were employed in training the models using the training dataset. ARIMA only requires prior data to generalize forecasts and model accuracy can be improved while maintaining a minimum number of parameters. Facebook Prophet facilitates easy and fast training of time series models as it requires less feature engineering. ES gives more consideration to the most recent observations in the time series hence usually producing accurate forecasts. The optimal order of the ARIMA models was automatically obtained using the pmdarima Python library as ARIMA (5,0,4) and ARIMA (3,0,2) for the temperature and humidity models, respectively. The growth of the prophet models was set to linear as no saturation was observed in the data and the changepoint prior scale was set to 0.01. Daily seasonality and weekly seasonality were set to true while the seasonality mode was set to additive since a constant trend change was desired.

Evaluation

The trained models were evaluated on the test dataset and the performance for each algorithm was obtained in terms of MAPE, Mean Absolute Error (MAE), and Root Mean Squared Error (RMSE). The evaluation is summarized in Tables 3 and 4, Figs. 7 and 8.

Table 3 Comparison of the performance of the employed algorithms for the temperature models

Metric	Facebook prophet	ARIMA	ES
MAPE	5.771	6.459	6.495
MAE	1.480	1.667	1.672
RMSE	1.819	2.041	2.037

Table 4 Comparison of the performance of the employed algorithms for the humidity models

Metric	Facebook prophet	ARIMA	ES
MAPE	8.981	10.294	10.160
MAE	4.199	4.635	4.594
RMSE	5.256	5.272	5.218

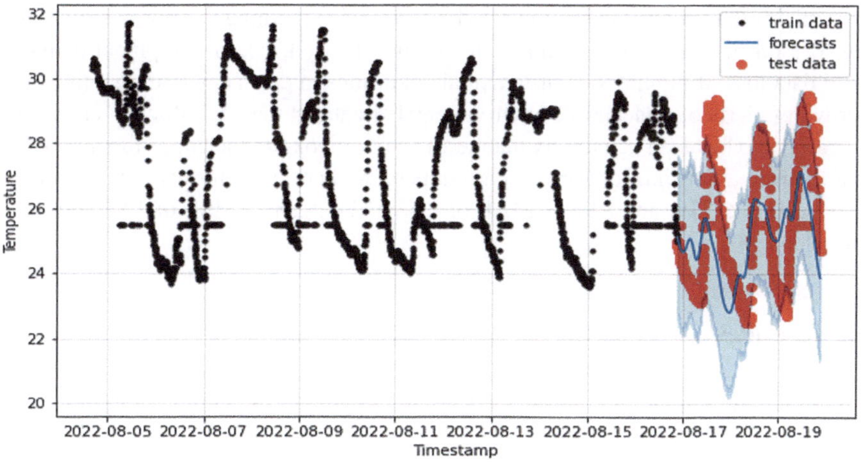

Fig. 7 Correlation of train data, forecasts, and test data for the temperature model

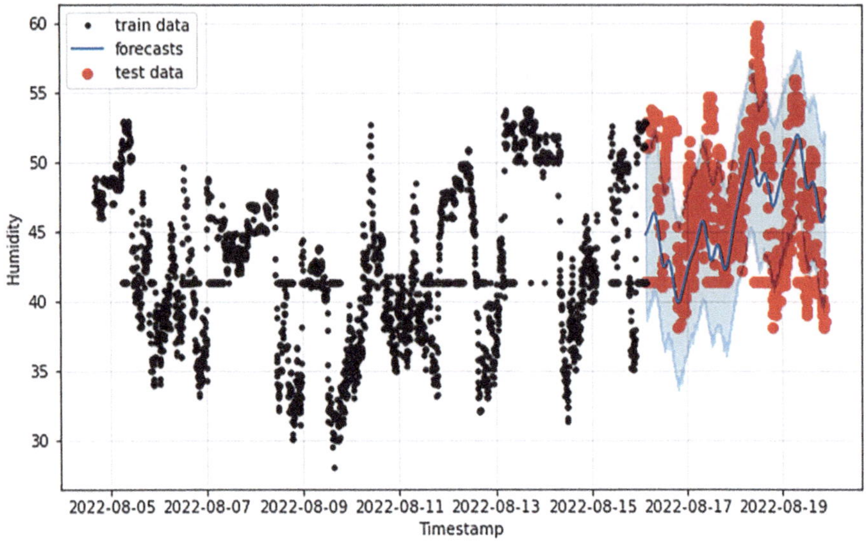

Fig. 8 Correlation of train data, forecasts, and test data for the humidity model

4 Conclusion

Data centers are an important aspect of many firms in the modern world facilitating their various business operations. Undesirable environmental conditions such as extreme temperatures, humidity, fire, voltage spikes and surges, and water leaks can lead to a decline in performance, unexpected failures, and total damage to equipment in the data centers. In this paper, we have presented an environmental monitoring system for data centers that utilizes IoT and machine learning to improve reliability as a result of decreased unplanned downtime, reduced risk of equipment damage, and data loss. It monitors environmental conditions in data centers in real time and provides alerts of undesirable environmental situations thereby allowing corrective measures to be taken in a timely manner. The Facebook prophet algorithm proved to be the best algorithm for forecasting temperature and humidity patterns, returning a MAPE of 5.77% and 8.98%, MAE of 1.48 and 4.2, and RMSE of 1.82 and 5.26, respectively.

References

1. Rasool, H., Rasool, A., Rasool, U., Raza, A., & Ahmad, W. (2014). Centralized environment and battery monitoring system for server rooms. https://doi.org/10.1109/ITEC-AP.2014.6941035.
2. Marzuki, M. J., & Newell, G. (2019). The emergence of data centres as an innovative alternative property sector. *Journal of Property Investment and Finance, 37*(2), 140–152. https://doi.org/10.1108/JPIF-08-2018-0064

3. Onibonoje, M. O., Bokoro, P. N., Nwulu, N. I., & Gbadamosi, S. L. (2019). An IoT-based approach to real-time conditioning and control in a server room. https://doi.org/10.1109/IDAP.2019.8875880.
4. Kurniawan, D. E., Iqbal, M., Friadi, J., Borman, R. I., & Rinaldi, R. (2019). Smart monitoring temperature and humidity of the room server using raspberry pi and Whatsapp notifications. *Journal of Physics: Conference Series, 1351*(1), 012006. https://doi.org/10.1088/1742-6596/1351/1/012006
5. Roy, A. et al. (2016). Energy-efficient data centers and smart temperature control system with IoT sensing. https://doi.org/10.1109/IEMCON.2016.7746251.
6. Kaliyamurthie, K. P., Nalini, C., Priya, N., & Geo, A. V. A. (2019). Smart server IOT based temperature monitoring system. *International Journal of Innovative Technology and Exploring Engineering, 8*(9S3). https://doi.org/10.35940/ijitee.I3118.0789S319
7. Yumang, A. N., Paglinawan, C. C., Sejera, M. M., Lazam, A. S., Pagtakhan, J. C., & Santos, J. S. B., (2017). ZigBee based monitoring of temperature and humidity of server rooms using thermal imaging. https://doi.org/10.1109/ICCSCE.2016.7893616.
8. Khan, W. Z., Aalsalem, M. Y., Zangoti, H. M., Zahid, M., & Afzal, M. K. (2018). Internet of things based physical and environmental monitoring system for data centers. https://doi.org/10.1109/ICRAMET.2018.8683913.
9. Nasution, T. H., Muchtar, M. A., Seniman, S., & Siregar, I. (2019). Monitoring temperature and humidity of server room using Lattepanda and ThingSpeak. *Journal of Physics: Conference Series, 1235*(1), 012068. https://doi.org/10.1088/1742-6596/1235/1/012068
10. Yamanoue, T. (2020). Monitoring of servers and server rooms by iot system that can configure and control its terminal sensors behind a nat using a wiki page on the internet. *Journal of Information Processing, 28*, 204–213. https://doi.org/10.2197/ipsjjip.28.204
11. Standards and Guidelines. Ashrae.org, [Online]. Retrieved June 13, 2022, from https://www.ashrae.org/technical-resources/standards-and-guidelines

GPSR Compliance

The European Union's (EU) General Product Safety Regulation (GPSR) is a set of rules that requires consumer products to be safe and our obligations to ensure this.

If you have any concerns about our products, you can contact us on ProductSafety@springernature.com

In case Publisher is established outside the EU, the EU authorized representative is:

Springer Nature Customer Service Center GmbH
Europaplatz 3
69115 Heidelberg, Germany

Batch number: 10761599

Printed by Printforce, the Netherlands